AF420690

Co-Science back to Life

Critics to the bio-synthetic anti-Nature delirium

Paolo Renati

© 2024 Obelisk Books – part of Obelisk Media B.V., The Netherlands

ISBN: 978-94-6461-133-5

info@obeliskbooks.com

Co-Science back to Life | Paolo Renati

Critics to the bio-synthetic anti-Nature delirium

First print: April 2024

Original title: **Miss Inco Scienza, ancora gravida?**
Al di là della tecnica: bio-co-scienze non integrate

Any part of this book may only be reproduced, stored in a retrieval system and/or transmitted in any form, by print, photo print, microfilm, recording, or other means, chemical, electronic, or mechanical, with the written permission of the publisher.

Contents

It is not so much the atomic bomb, of which so much is spoken, to constitute, as a deadly device, the mortiferous. What has long threatened man with death – and with a death that concerns his very essence – is the unconditionality of pure will, in the sense of deliberate and global self-imposition. What threatens man in his very essence is the deceptive belief that, through production, transformation, accumulation and governance of natural energies, man can make the human situation easy for all and generally happy. But this peacefulness is nothing more than the uninterrupted agitation of its most unrestrained self-imposition, now oriented upon itself. What threatens man in his essence is the belief that the realisation of absolute production can take place without any danger. [...] What threatens man in his essence is the conviction that technical production will 'put the world in order'; whereas, on the contrary, this kind of order levels every *ordo*, that is, every *rank*, in the uniformity of production, thus dissolving, from the start, the possible origin of every rank and every recognition from the foundation of *being*... it's the same technology (*techne*) that blocks the understanding of its own essence more than anything else.

Martin Heidegger, *Why the Poets?* pp. 271-272,
in *Interrupted Paths*, 1949

Introductive note by the author: comment about the historical present

It is good to know that this text was drafted in its first version in 2017, undergoing some revisions along 2018, the year in which it reached its final form. Publication only became possible three years after its completion and after no less than six editorial rejections (five of which were shoot without any justification, but when one reads this text, this will be understandable, almost taken for granted). This hindrance was accompanied by no small amount of torment and a sense of impotence, especially at the idea that what had to be shared and brought out was of extreme importance and urgency.

In fact, it was a bitter surprise to find that the very deceitful drifts towards which we were trying to place the reader in critical observation – to promote a resolute and urgent change of receptivity and anthropological course – were condensed into such dramatic and accelerated circumstances and recrudescence as those we are currently experiencing. We are, of course, referring to what is being referred to as the 'Covid19 pandemic': the greatest delirious mystification since the days when men became victims and executioners in cruel mass sacrifices in the name of some 'god' demanding 'debts' in terms of 'lives'. The archetype, it seems, has not extinct.

And, mind you: It is not a question of denying the suffering or deaths of anyone, but rather of debunking the narrative that boasts of their catastrophic exceptionality (first and foremost numerical), the (postulated) unequivocal and indisputable causal relationship between viruses and symptoms, as well as the unprecedented negligence in health care (in terms of the management of structures and personnel, the unworthy blackmail of the latter, collusion with a policy that should only have promoted life instead of destroying it in the false name of its defence). The matters concern the immense scientific hypocrisy such as that concerning the causes of death or the use of swabs and diagnostic tests, as well as the theories of contagion – even for asymptomatic people! – the touted usefulness of masks depriving facial expressiveness and social distancing, the idea that a 'vaccine' [indeed an experimental gene therapy!] is the only solution, etc.

Added to this there are the indecent and unspoken conflicts of interest between: healthcare companies, governments, pharmaceutical companies, hi-tech companies, and digital platforms (such as search engines, medial-contents sharing-sites and various social networks), as well as unscrupulous and age-old backroom lobbies (whose conflicts of interest are obvious even to a kid, by doing a little research on the web. Here there's much more than *conspiracy*!).

But the most tragic aspect, which makes the adjective *"delirious"* justified for this all, lays in having seen the ease with which an overwhelming majority of 'human beings' could replace truth with lies, common sense with paranoia, respect for the other with violent imposition on the other, replace civic sense with bovine obedience, the value of contact and feelings with terror, with disgust for the 'potentially infected' and with the denunciation of the 'criminal without a mask', adhering to and supporting with all their might such an immense spell (which has thus become a true catastrophe). This was possible because such a catastrophe (constructed on the way like a self-fulfilling prophecy) has been the subject of a compact, planetary-scale and compulsively repeated narration, all aimed at emotionally hooking the audience through the tragic nature of masterfully exalted facts, on which, however, no clarification has been provided about their real causes or contextualisation. In addition, every discussion and every piece of information that would have blatantly and easily compromised the only apparent coherence of such a mammoth and lethal antics was punctually removed from the public debate. Indeed, one only must ask oneself a few simple questions to see that more than 'something' does not add up at all.

In any case, the responsibility of so many individuals, constituting the believing and uncritical masses, who not only nurture such a narrative, but also defend it, cannot be ignored. If such "theatres", which are always run by an elite, can be established, and persist over time, it is because there is a more or less explicit consensus among the majorities, history teaches us. And, mind you, when we talk about the "masses" today, we are not talking, as in earlier times, about the uncivilised and illiterate, but rather about educated people, including engineers, scientists, biologists, doctors, lawyers, teachers, entrepreneurs, business executives, and so on, who actively contribute to building technical skills, disciplines, fields of research and developments for society as a whole. Those same technical skills that, as T.W. Adorno, might also say, have evidently been introjected in the solitude of specialisms or in the absence of a true *Bildung* (inner building) that can produce in each subject a *Weltanschauung* (world view) capable of giving an inner compass to our being humans and living systems.

And here it becomes clear how the role of School (in the broad sense) that does not promote *Bildung* in the young person today, in fact underwrites the competent and specialised automaton (already forecasted cyborg) of tomorrow. Indeed, who knows why - one wonders - the didactic and scholastic component has been precisely one of the most hardly 're-written' by this 'pandemic'? It seems that someone knows full well that, if one defunctionalizes at an early age certain process of physical, analogical and emotional contact, relating and learning, one paves the way for acquiring a subject who lacks certain tools of discernment and - it has to be said - of sensing the other (and oneself). These are the same *feelings* that seem to be missing from the majority of people who are capable of believing what is theatrically staged for them by the system's media interfaces.

When it is thought that those who do civil disobedience are doing it 'at the expense of others' and that with their 'irresponsible selfishness' they are putting their neighbour 'in danger', and nobody realises that even obedience (uncivil, as uncritical) cannot be done 'at the expense of others' (Hanna Arendt teaches us), one is the first to betray an authentic democracy, since there is a third way, respectful of every point of view, and it would be the one that allows those who feel in danger and want to make their own choices to cover their own faces with masks, distance themselves, lock themselves in their homes, vaccinate themselves, etc. to do so, just as those who do not wish to do so must be free not to. And the same must apply to any health treatment (vaccines above all). And, please, let's not say that those who do not follow certain orders (like who do not inject themselves with the transgenic serum) constitute a danger to others, because this is simply false and in obvious contradiction to the very principle for which "vaccination" is promoted: if one is vaccinated, why should he/her fear those who are not?

Does this bring into play the issue of public expenses and the medical care that would be needed by all those 'reckless' people who have not protected themselves and who would become "much more ill"? Apart from the fact that the meta-analyses on vaccination studies speak clearly about how much this is just a myth, secondly, then, smokers should be prevented from smoking, everyone should be required to do two hours of physical activity a day, to eat five portions of fruit and vegetables (because the 'scientific publications' say so!), certain foods should be abolished, one should not take walks on bumpy paths, competitive or extreme sports, explore, risk anything... one should be prevented from living life!

But when will people understand that the issue of 'safety' can never subordinate that of one's own intimate freedom, at least that of managing one's own body (a sacred and inviolable temple!) and taking responsibility for what one wants to experience? If there is no this, there is no life, but only a pre-packaged surrogate.

It is then strange that all this 'precautionary nature', however, is not also promoted with the same rigidity with telecommunication networks (see 5G and co.), with pesticides in intensive agriculture, with nuclear weapons to be kept in the army bases, with genetic editing and GMOs, with toxic chemotherapies...and so on, in a long list. Smell of conspiracy or, perhaps better, 'let's stop kidding ourselves' ...?

As will become clear in the course of this booklet, it is high time that every individual took to heart, as the first aim of his or her life, to conduct the most refined self-education possible for himself or herself and for others, and that he or she matured an awareness that would allow an emancipated vision with respect to those subjects of history (with Günther Anders) who in this modern age are replacing the human being, seducing him or her towards an identitarian, controlling and a-biological attitude: technology, capital, bureaucracy.

- When one does not see biologists and doctors loudly, and *en masse*, admonishing for the scientific absurdity taking place and not inviting everyone to build up their own immunity to a (presumed, since never really isolated) virus in the way it should always be done – i.e., by letting it be encountered and processed by biological (human) systems that at 95% do not develop symptoms and at 5% have at their disposal highly effective therapies (which, if cleared and approved, would not even require hospitalisation) – it means that something is wrong.

- When people do not understand that a face mask constitutes an absurd obstacle to free breathing and that its 2 micrometre intra-fibre spacings (not counting lateral spaces) could never stop a 100 nanometre particle, and when they do not understand that, if there really were a lethal virus running around, no face mask would save anything or anyone and we would be looking at a catastrophe (but a real one!), and when they do not understand that, in the aforementioned uselessness, such a mask becomes an unacceptable removal of a hinge (our face) that is essential to the neurobiology of the human being (not to mention children!), it means that there is a lot wrong.

- When a health care system does not clear effective and inexpensive therapies (such as large auto-emo infusion of oxygen/ozone, hydroxychloroquine in conjunction with azithromycin and low molecular weight heparins, corticosteroids, high doses of vitamins C and D, or hyperimmune plasma), and instead places people, after days of administering paracetamol, in intensive care with forced ventilation, it means that there is something big wrong.

- When biologists and geneticists do not raise loudly about the absurdity of using an RT-PCR swab that cannot tell to whom or what a gene sequence (amplified by a primer) belongs, while there are about a hundred primers for Sars-Cov-2 in the world and therefore no 'gold standard' exists, and after it has always been recommended not to use this type of test for diagnostic purposes, and when even a table, a goat and a papaya fruit (and even a packaged apple puree) resulted as "positive", and when one speaks of 'waves' based on the results of such a test without assessing how many people really have symptoms (which would constitute real 'cases', without taking into account that seasonal influenza every year makes about 5-6 million symptomatic patients, and 15-18 thousand deaths that have always been reported in national newspapers, without sending a country into panic, lockdown and collapse), then it means that there is even more wrong.

- With such a percentage of asymptomatic people, even assuming that the RT-PCR swab were reliable, even a brain-damaged person would realise that the increase in the number of positives would be nothing less than wonderful news, as it testifies that the population is increasingly capable of being in contact with the virus and not developing disease; and this would make one realise that the simplest, healthiest and cheapest method of achieving the mythological 'herd immunity' would be not to vaccinate, but to let the circulation of the 'new pathogen' take its course (since it has very high a-symptomaticity rates and for those with already problematic conditions – the only ones really possibly at risk – there would be all the therapeutic tools to manage the course of the body's response).

- But when, at the end of 2020, one does not notice that the 36,000 deaths 'due to Covid19' (or thereabouts), to be added to the average number of annual deaths in Italy (about 648,000), are not there; and when one does not realise that, therefore, 'labelling' games have been played - since the other causes of death have not disappeared - and when one does not take into account that each hospital received €5,000 to €9,000 a day for each patient labelled /by the fake swab) as 'Covid19', then there is much more that is wrong. And the list could go on.

- But the most dramatic point is that when bureaus force children and teachers to wear (unnecessary) masks and the mothers of a country do not all unite under a healthy feeling of indignation and maternal protection of their children from hallucinating distortions and gratuitous violence, the effects of which are potentially dramatic in the long term, well... then there is really, really a lot wrong.

What is denounced and discussed in this essay/manifesto is precisely that cultural substratum that makes contingencies such as the 'Covid19 pandemic' feasible, lasting and, tragically, repeatable. Yet this work makes not a single mention of such a historical circumstance, as it was completed at least two years before it occurred. It does not speak of any pandemic, but of all that constitutes a premise even for such a circumstance and of all that, through the deliberate management and narration of the latter, would be irretrievably lost, unless we mature a profound review on our *being in the world* beyond respectability, 'security', data, and 'scientific' descriptions.

We therefore wish to point out the extent to which the themes that come together at a rapid pace in this booklet are all too often considered separately (especially with respect to events such as a so-called 'pandemic'), when, on the contrary, the 'removed background' – inhabiting the less and less human social context of industrial, capitalist, technological, scientistic, narcissistic, respectability-culture, with the taboo of death and (therefore) trans-humanist – constitutes precisely that terrain on which the *hybris* takes root, which, in extreme synthesis, have as their common denominator the estrangement from the biological *logos* and from *Physis* (Nature) meant as a *womb* to which to reserve the utmost respect and within which to put in action a human *modus* capable of a fully empathic and authentically perceptive relationship with what exists.

Paolo Renati
January 2024

Foreword

If we want to give to our vital habitat, which is so wonderfully rich and intelligent, and consequently also our species, the concrete possibility of a harmonious and flourishing existence, it is necessary to conduct a reflection on how, how much and why *technology* has now become an *environment* and not just an *instrument* of the human being. It is no longer enough to acknowledge that it was the attitude through which implement *actions* aimed at purposes not possible to be fulfilled within the options already present and 'given by nature', but it must be recognised that the results of such action have changed the premises of action itself and of conceiving ourselves as 'humans'. Such *oriented action* is in fact only the mirror of a psyche that, by its very nature and history, acts an analysis of reality and formulates hypotheses, i.e., that connotes a continuous circular wave between objectification and subjectification. The questions are: beyond cultures, ideologies, technical achievements, beliefs, or various identifications, does our current thinking of ourselves as human beings still have a chance to be deemed *human*? Does what we consider normal, right, and appropriate for our contemporary being in the world have a real biological *grounding*, a real biological *plausibility*? Why should we ask ourselves this? How is *technology* relating to *nature* and what are the effects of this?

If, in social, environmental and biological as well as in economic, political and cultural terms, it is evident (hopefully to everyone) that the current human condition, and that of the planet, lays in an obscure state, then it goes without saying that we need to address the (primarily *existential*) premises for technical action (which today is expressed as much in the synthesis of neo-materials as in the gestures of bureaucracy, in interventions in the landscape as in the training plans of public or private education). With "existential premises to technology" and to the gestures that spring from it, we are questioning ourselves about the presence (very important, vital) of a *feeling*, tender and great, for life, for its spontaneous declination in nature and for that subtle network of irreducible connections with which it is connoted. That feeling that calls for a profound and delicate care that, especially today, seems not to find due expression in the human way.

Where is that *feeling* today in the *homo technicus*? With its innumerable alienations (first and foremost those of synthetic biology and the various distortions pertaining to the generation of life), today's *technical environment* increasingly appears to be the manifestation of the suffocation of that feeling of care and relationship for a *quid* ('something') *de facto* sacred because prior to, and premise of, humans themselves. Such a sentiment, based on the recognition of the *sacredness* of what is produced through the autopoiesis of life in maintaining the connections of everything with everything, is vital and necessary because it allows each subject to always relate his/her own actions to what is profoundly important, since inescapable for the existence, and therefore *untouchable*. Recently and increasingly, this sensibility, which has always rooted every human being in the sacred communion with his own vital womb (*Nature*) seems, tragically, to have atrophied not only in the gestures of *big capital*, but above all in those of research and science, as well as in the attitude of the 'average citizen' who seems to increasingly seek answers to his needs in technology. Needs that, however, often narcissistic, are induced precisely by technology itself. And therefore, the dichotomy between *nature* and *culture* increasingly sinks its cut into a human heart more and more identified with the latter, but bleeding because of the offence to the former.

Looking at it in its entirety, the *body* of argumentation here moves from reflections on the current state of the art of that (simplistically) so-called 'biotechnological' ambit and, inevitably, also points to the most crucial issues of bioethics. All this has emerged, almost involuntarily compared to an initial intention, in which considerations (felt to be urgent and due) on some recently published research work in the field of genetics (see below) were followed by inevitable meditations on what *life* is, what *Nature* is, how, how much and why the human being is *of* Nature and *in* Nature, of the relationship between *technology* and *biosis* that constitute the human and, therefore, also of the relationship between the (super)structures of *logical* thinking, the one whose main aptitude is objectification and division into parts, and the innate intrinsic features of *analogical* feeling, which typifies the unrepeatable subjectification.

The former is the premise of the technical gesture and is constitutively alienating insofar as it is dual, producing an observation from the outside and the identification of causes, distinguishable from effects, within an arrow of time in which the *before* and the *after* (as well as the *here* and the *there*) accommodate distinct 'portions of reality'. The second is the womb of the process of life insofar as process of perception[1], of adaptation (in the first instance thermodynamic) to the conditions of an

---

[1] For this, it's enough to see the entire phenomenological tradition, from Bergson to Husserl to Merleau-Ponty (see recommended works in the bibliography).

environment, the *qualitates* of which also come to acquire meaning when some subsystem tends to want to preserve themselves (as a living beings). Such perception cannot be dual, since the perceiving of an external environment or an internal state is only and precisely in the perceiver's change of state, i.e.,, the perceiving subject is the object of its own perceiving of any object. Otherwise, this would not be *feeling*, but only thinking. It is within feeling that the relationship we are narrating lives, which, analogically (since not dually and without autonomy or arbitrary isolation of *parts*), truly subjectifies a living being (such as man) in the environment insofar as it makes him/her participatory beyond any conscious observation, judgement, or action.

The level we are interested in moving to is the latter, since it is the one that is ontologically necessary and that also accommodates (and overcome) also what is produced by logical thinking. We feel that the problem of technology has not yet found an adequate solution, since it is only addressed within the level in which technology already has full domicile, while the problem of its premises can only be addressed there in that greater *self*, in which 'only then' the "I" self-produces and represents itself.

These themes, let it be clear, are addressed from a perspective without any citizenhood, not ascribable to any strand of thought, nor to any affiliation. Except for the only one proper to every living being: that of the *logos* of Nature, the bio-logos, the *bio-logy*, the adherence to which is certainly not a choice, but a factual condition. What is set forth here is therefore intended as a non-opinion; no ideological position is being expressed, but rather the intention is to communicate, to make visible, to show the very simple truthfulness that is inherent in Nature and that is im-mediately delivered by it, provided that we decondition our gaze, which is deformed by the aims and projections with which modern humans now represent themselves and, likewise, relate to their own biological essence.

And whoever, reading this, could find himself/herself personally challenged (due to his/her own condition, his/her own ideas, his/her own desires) – we heartily recommend – let him/her pause for a moment over a possible *motus* of irritation, resentment, or rejection, and try to listen to what it is that is being moved and what makes him/her feel 'touched'. Please, consider what is written here for what it *is meant to be*, as the voice of an authentic condition that comes from far away, without any judgement or vexatious intent, but rather inhabited by the breath of an urgency that is anything but superfluous.

Some aporias are inhabited by all of us. And perhaps the call to a dialectical thought is realised precisely in not claiming any resolution of contradiction, in absolute terms. Let us, however, at least critically address what is manifested in front of us, with rootedness, i.e.,, keeping in mind what we are and what we cannot disregard in order to truly *live*, instead of only benefiting from *existential surrogates*. Otherwise, a suffering so subtle, yet ubiquitous, as to remain faceless and without comprehension, so that the individual's wriggling in the a-biological horror would demolish both the subject and his own environment from which he cannot think of himself/herself as something *else*.

"Researchers, before they become researchers, should become philosophers. They should look at what the purpose of the human being is, what it is that humanity should create. Doctors should determine at a fundamental level what it is that human beings depend on for their lives"

(Masanobu Fukuoka, *The Straw Thread Revolution*, Ch. 15: *Limits of the Scientific Method*, p. 97, 1975)

Introduction

Understanding that the problems of modern society are the direct expression of the problems of gnoseology, and epistemology would be a great achievement for mankind and perhaps the only real hope of survival for the entire planet. It may seem paradoxical, and to most even sacrilegious, but the very *forma mentis* that has underpinned the development of science, right up to its current shape, constitutes the very premise that has led to the problems that science itself believes to be able to solve in the way it is attempting, without realising that in doing so it is merely subscribing to the only real problem: the dual and dichotomous way in which we relate to existence and therefore to knowledge. Although the mindset of many intellectuals prevents *ab initio* the possibility of a *vision* such as the one we are suggesting, we nonetheless venture to divulge these reflections, hoping to be somehow supported by the fluctuations (perhaps not so stochastic) of what some call *deterministic chaos*.

If until a little over half a century ago, it was still legitimate to have doubts about the above, this is no longer the case. To testify this there is the ever-increasing number of research results and what transpire to be the 'underlying beliefs' that are the prerequisite to those kinds of 'scientific results' and are continually reinvigorated by them. This extends from particle physics to genetics, from physical chemistry to medicine and neuroscience.

The methodological criticality that afflicts both the quantitative and qualitative growth of the corpus of human knowledge lies not (only) in making research that is only verbally aware - but factually unaware - of the false coincidence of description and reality, but rather in the implicit postulate that the latter consists substantively of possible parts within itself. That is, it is taken for granted that reality is 'made up of parts that exist in themselves'. This position from which to start any analysis prevents any *relational* vision of reality[2], in the sense that within any description one wishes to undertake, no matter how complex and comprehensive, one will always have to commit an artefact depending on where one has decided, or should have decided, to end the range of observation. The 'dramatic' aspect is that it seems that it is no longer possible to produce knowledge except through description.

---

[2] *Physical Analogical Foundations of Conscious Reality*, Paolo Renati, *Analogical Con-Science Series • 2*, ÌNIN Editions, Lugano, October 2016. ISBN 978-88-98497-09-6.

This is true because 'knowledge' that is not subordinated to description would not be expendable into objective, transversal, and repeatable terms. These are all connotations that define any practice involving collectivity and, like technology, are subject to every form of interaction and exchange (including legal and monetary). In fact, it is through affinity of views that the twinning of science and capital, and later, in the current phase of *absolute* or post-bourgeois *capitalism*, and even the subordination of the former to the latter, was historicised very early on. Robert Musil already grasped well the correspondence of this 'common evil' to human expressions – which lost the *analogical* quality along the way (we would say) – who in chapter 72 of *The Man Without Qualities* makes Ulrich reflect in an inner stream of consciousness as follows:

«If one asks oneself without prejudice how science has taken on its present appearance – which is important in itself, because science reigns over us and not even an illiterate person is saved from its dominion, since he learns to live with innumerable things that were born erudite – one gets a rather different picture [from the one about grinning under the moustache of the scientists mentioned previously in the chapter of Musil's book, *ed.*]. According to reliable traditions, it began in the 16th century, a period of very strong spiritual movement, to no longer strive to penetrate the secrets of nature, as had been the case in two millennia of religious and philosophical speculation, but to be content with exploring its surface, in a way that cannot help but be described as superficial. The great Galileo Galilei, for example, the first name that is always mentioned in this regard, got rid of the problem of what is the intrinsic cause of nature's horror of empty spaces so as to force a falling body to traverse space after space until it reaches solid ground; and contented himself with a much more vulgar observation: he simply established the speed of that falling body, the path it travels, the time it takes, and the acceleration of the fall. The Catholic Church made a grave mistake by threatening such a man with death and forcing him to recant instead of killing him unceremoniously; for him and his peers' way of looking at things has – in a very short time, if we use the measures of history – given rise to the railway timetables, the machine tools, the physiological psychology and the moral corruption of the present time, and he can no longer remedy it. The Church probably made this mistake out of too much caution, since Galileo was not only the discoverer of the motion of the earth and the law of falling bodies, but was also an inventor who the big business was interested in; and moreover, he was not the only one who was pervaded at that time by the new spirit; on the contrary, history teaches us how the cold positivism, that animated him, spread as freely and untamed as an epidemic, and as shocking as it may be to hear that one was 'animated by cold positivism', since we can say that there is even too much of it, at that time the awakening from the slumber of metaphysics, to stick to the strict examination of things on the basis of the most varied testimonies, must even have been an intoxication, a fire of positivity! But if the question arises as to why it occurred to mankind to change in this way, here is the answer: mankind simply did what all sensible children who have tried to walk too soon do; it sat down on the ground and touched it with a not very noble but safe part of the body, let us say: with that part on which one sits. The strange thing is that the earth showed itself so sensitive to that contact, so as to allow itself to let knowledges discovery and comfort be torn away from it in an abundance that has something miraculous». [...]

«And indeed the passion for facts, before intellectuals discovered it, was only possessed by warriors, hunters and merchants, that is, by witty and violent beings. In the struggle for life there is no speculative sentimentality, but only the desire to eliminate the opponent in the quickest and most effective way, then we are all positivists; and similarly in business it would not be a virtue to be cheated instead of playing it safe, where profit ultimately means a psychological overpowering of one's neighbour brought about by favourable circumstances. But if we consider what are the qualities that lead to breakthroughs, we will find freedom from the usual scruples and inhibitions, courage and just as much resourcefulness, the will to destroy, the ability to disregard moral considerations, a patient bargaining for the slightest advantage, the stubborn expectation, if necessary, that the goal will come true, and the cult of measure and number, which is the most fitting expression of distrust of everything that is not certain; in other words, we will find nothing more than the ancient vices of hunters, soldiers and merchants, simply transferred to an intellectual plane and converted into virtues. In this way these vices are taken away – this is true - from the spasmodic and rather vulgar pursuit of personal advantage; but not even, as a result of this transformation, it has been lost what could be called 'the element of the original evil', since it seems indestructible and eternal, at least as eternal as all human greatnesses, consisting in nothing else but the desire to trip up such greatnesses and to see them thumb their noses»[3].

---

[3] *Der Mann ohne Eigenschaften*, Italian translation by Irene Castiglia, *L'uomo senza qualità*, Robert Musil, Second Edition, February 2016, New Compton Editor; Author's English translation from Italian.

It should be noted that this is not an 'attack on science' *tout court*, and even less to Galilei, quite the contrary. Rather, it is a subtle, ironic, and indispensable (constructive) criticism of what true science is, indeed, plagued and sabotaged by. That is, the same plague that degrades *communities* to societies of norm, procedure, and currency exchange. That is: the technical and calculating attitude that, at least for naivety, no longer grasps what dualizing approximations have articulated the descriptive *logos*. In this way, the clear consciousness of *being within* what is being investigated, i.e.,, reality, and not beyond it, becomes increasingly evanescent, together with the strengthening of the idea that reality is a summation of distinguishable parts (and observables), with which is associated the progressive substitution of what is real with what is described/represented.

To give an example, we all believe that the colour of an object, or its net electric charge, are independent of its mass or of its surface conformation, or of the mechanical properties of its material, so we identify physical *observables* such as angular momentum, mass, electric charge, space, and/or logical and phenomenological observables such as colour, quantity, type, etc., as absolutely independent. That is, one can consider a red, smooth, plastic ball of mass 20 g, at rest, here, electrostatically charged: or the 'same' ball, electrostatically neutral, yellow, rough, in accelerated motion, over there, etc. One would thus deduce that the variables-characteristics of the system are mutually isolable: one can change mass independently of colour, electric charge independently of roughness, or volume... or cannot? One must be careful: this is only true up to a certain scale of observation! For example, at fundamental levels, the oscillators that make up the chromatic groups of the polymer of the little ball (which macroscopically give it a colour depending on which frequencies they absorb/emit) are not independent of their electrical charge distributions, of their oscillating masses that decide which frequencies they absorb/emit (colour), and so they are not independent either of the shape of the molecules and therefore of the surface that the material 'coloured in that way' will have. Therefore, even the surface 'finish' (i.e.,, 'roughness'), at the atomic and orbital scales, is not independent of colour. Moreover, even the colour visible macroscopically from the outside is not an intrinsic 'property', since it depends on what kind of light hits the body and what range of the electromagnetic spectrum it picks up and may also depend on relative velocity vectors associated with that object; and the net electrostatic surface charge also decides how the molecular groups are polarised, and thus, intimately, also the surface topography, its roughness, its tribological (frictional) properties. And all this also depends on environmental conditions such as temperature, humidity, fields and more. Furthermore, is the mass of that object something that is very independent of the relativistic effects associated with the energy size of photons, giving that object that perceptible 'colour' when that mass is hit by light or not? To what extent are the net electric charge and its electrodynamic effects, dependent on the state of motion, unaffecting on the 'colour' or mass of that body? Or how far can the mass and electric charge of, say, an electron be decoupled? And so on.

Surely, almost anyone with a good 'scientific background' would argue that these are illogical inferences or mental extravagance, even inadequately placed since such considerations are reasonable if the energies, velocities and dimensions involved justify it (in fact, for example, quantum indeterminacy about the velocity in m/s of a pencil placed on a table assumes non-zero values only if one goes to the 25th digit after the decimal point!). This is true for the calculations and phenomenally, nothing to say. But it is not true ontologically, and the fact that an uncertainty is indelible in every observable per se and implicit in every act of measurement, in fact, expresses that original indivisibility that is corrupted in the act of description that needs to identify parts[4].

Moreover, this interdependence of properties, which in physics become true observables (electric charge, mass, momentum, position, etc.), also called *proximality of the properties* in neuroscience, is a decisive aspect in experiential and perceptual processes (i.e., in the realm of the living, of the bios) and highlights how considering the various properties, the various physical observables (and, if you like, the various identified *degrees of freedom* of the system) as independent and "existable" in themselves, is a big mistake. Take, for example, *Benham's disc* (a game

---

[4] Renati P., *op. cit.*; regarding the irreducible approximation of knowing by means of a finite and local observation, it is also interesting to recall the original correlation *consciousness-world* in Husserl (*Ideas for a Pure Phenomenology and for a Phenomenological Philosophy*, Book I[st], sect. IInd, chap. I[st]), in which the identification of parts taken for granted, by the natural sciences and everyday experience, is revealed as the primary obstacle to achieving the *essence* of reality; In contrast to this, *epochè* (the suspension of judgement, the phenomenological attitude in opposition to, and overcoming, the natural one) is promoted, which, different from the *destruens* 'sceptical doubt', is instead *construens* insofar as it aims at the attainment of the essence of reality, recognising that it can only proceed with reference to subjectivity, without however denying an existence in itself of reality (*«the world is already there»* and becoming). Paradoxically, in the obvious *givenness* of the natural stance that observes reality through parts, layers, intervals (thought of in isolation due to the need to manage the datum), one gets bogged down in a vision much more affected by beliefs than the phenomenological approach can produce, which in the process of knowledge construction does not shy away from uncomfortable subjectivity and does not take anything for granted, especially the modus by which we observe.

constructed by Charles Benham in 1894 that is in fact a painted spinning top with a white background on which are affixed short black stripes shaped as concentric arcs at various distances from the axis of rotation). When the spinning top spins, pale colours emerge and their tone (red, blue...) depends on the speed of rotation (and possibly, therefore, on the radial distance from the centre).

If we overcome the false preconception that they are only 'mental colours', because we have already integrated the fact that an experience is *always* a physical[5] process because it happens in nature (i.e.,: in *physis*), a Maya's veil falls, hiding from us the extent to which the idea of *qualitates*, properties, degrees of freedom, etc., identifiable in a system and thinkable of as autonomous categories (in Aristotelian way), is at the very least an arbitrary and even erroneous postulate. Furthermore, the idea that *qualitates* can be attributed to objects that should exist noumenally in themselves independently of the former (facts of a *substantia absoluta*) also dissolves. And even the (Platonic) idea of a contingent world in which *becoming* renders things approximate to a hypostatised *being* (metaphysical *ideas* eternally yearned for and unattainable) collapses, insofar as one grasps the profound (physical!) rooting of being in the becoming *òlos* constituting itself of continuous self-relationship[6]. This is a profound problem of scientific knowledge, which paradoxically, ends up believing *representations* that are the result of postulates that have not been noticed (and questioned) or thought of in depth.

It is usually said that what is not detectable is not verifiable, hence cannot be considered within an experiment, i.e.,, it is not part of 'science'. Unfortunately, the step from 'I cannot measure it' to 'it does not exist' is very short. And consequently, the step from 'science = dealing with what exists' to 'science = dealing with what is measured and reproduced' is even shorter.

It would take a clear realisation that, although it is not measurable, it is untenable that, in reality, the 'individuated parts' are not, after all, connected at that level of intimacy and that they are not, in truth, the *result* of

---

[5] See the research relevant to the theoretical perspective of the *spread mind*. Despite some conceptual gaps on the idea of 'body' and the inconsistent postulated identity between the reality of an object and the perception of the same (indeed, this, a sub-set of the former), a stimulating reading is: (tr. *The Spread Mind - Why consciousness and world are the same thing*), *La mente allargata - perché la coscienza e il mondo sono la stessa cosa*, Riccardo Manzotti, Ed. Il Saggiatore, 2017 (Italian).

[6] See Renati P., op. cit. and pp. 33-35 of *First elements for the foundation of a new paradigm in physics*, Paolo Renati, World Futures - The Journal of New Paradigm Research, 00: 1-22, 2015 Taylor & Francis Group - Routledge (August 2015) DOI:10.1080/02604027.2015.1018054

those ubiquitous *relations*[7]. Although all of this, technically speaking, does not change much in the practical-immediate act, it would allow us to always keep in mind the degree of connection existing in reality with respect to the fragmentation returned to us by the represented/described, and therefore the measured, level. Even though the degree of fragmentation and resolution, in fact, resolves ever smaller and smaller 'mosaic tiles' – through the descriptive vision that identifies independent postulated parts and observables – who and how can care about the connectedness that underlies these described tiles/parts?

As will be seen, this may not be relevant in the study of inanimate systems but becomes crucial when one begins to deal with biological systems. It is precisely to the effects of this problem that we wish to address in this paper. And, as will become clearer, the epistemic removal that portrays a reality understood as 'manipulable through the management of "pieces"' is accompanied by an existential drama in which the 'pieces' have become the subjects and 'parts' of themselves. This is how the *heterogenesis of aims* takes place in which the *general equivalents* of the collectivised social setting (such as money, law, education, healthcare, the citizen, the market, science, language) are passed off as *ontological invariants* (i.e., existing fundamentally, *per se*) and disguised as the 'existential requirements' of human 'being in the social world'. When, on the contrary, those *forms* have become positivised from an identitarian position that has been unable (unwilling) to go beyond the alienated phase of the *describing subject*. And thus, such heteronomous structures appear as transcendent to the individuals who are born into them, but falsely; and that "veil of Maya" will resist until the subjects (i.e., all of us) rediscover its strictly ideological connotations, and therefore determinable, rewritable by them. Until such rewriting takes place, the social system to which technical and economic man has given rise, and which constitutes his 'second nature' (with Adorno), will be the only true *aim* (law for law's sake, progress for progress's sake, technical power for technical power, markets for markets, etc.) and the human being, together with his biological world, will be the only *means*.

Therefore, we need anything but encouraging an interest in science among young people by extolling the practical aspects. What is needed is the exact opposite: if we want to educate a generation of future scientists, researchers and academics capable of promoting such redemption, we must re-evaluate the importance of *theory* and reflection freed from the

---

[7] Renati P., op. cit.

expendability of *experimentum*, which realizes *first* the *categories* by which the investigation is moved, the simplifications from which it is affected, so that the results, albeit approximate, are spent in the real awareness of what they represent. To do this, the relationships between genuine research, visibility of publications and allocation of funds must be totally rewritten, redeeming them from the logic of capital or prestige, allowing for an honest pluralism that brings the voice of currents of thought that also challenge the current technocratic assets. Philosophy disappearing in the coils of praxis or withered on the pavement of mere logics must be recovered. Indeed, reflective thought must not lose the 'spiritual' and *sensory* component that pertains to the profound connection with the body, life, nature, and feelings.

The anthropological course of the human species (at least for that vast majority of it which has been relying on *techne* without a spontaneous *demeanour* that would allow the preservation of its *bios* unlike certain 'tribal' niches have been able to do) is in essence the expression of a hermeneutic cycle that has seen the beginning of the alienated phase from the first objectifying gesture, then from the first descriptive and instrumental acts, then: with language and with technology. Inevitable. But every alienated phase (the *antithesis*, in the case of the "second nature" of the techno-social system), in order for it not to turn out to be self-destructive – implying that men spend their existences imitating and bending nature from the outside instead of living it and inhabiting it – requires a *synthesis*, a recomposition that envisages its own reintegration into the *thesis* (the undifferentiated man-nature) on a self-conscious level. The time available to accomplish this before an irreparable biosphere *collapse* may be too short.

To those who want to see, agony becomes inevitable due to the delirious super-multiplication of habits, goals, global decisions, trends that have in common only the further underwriting of the gap between two poles. The first concerns what that biologically subjectifies and roots man in a healthy and phylogenetically evolved *analogical relationship*; the second concerns what that, seductively and technologically, withers the individual now identified in a cybernetic representation, the fruit of an invading *descriptive communication*, nourished by calculation, exploitation, and repetition. The dichotomy is not here set *a priori*, it is not therefore a Luddism-like stance; it is quite clear that there could be a harmonious composition between a *technical ego* and a *biological self*. The problem lies in the proportions acted and actually allowed. When the life of humans (of the so-called 'affluent' part of humanity) is sucked for the 80%

in time into a role that needs to be 'played' for the guarantee of survival (i.e.,, their job), it is clear that 'biological existence' remains an empty definition since the horizon of experience is all codified and controlled within what the system (the alienated and alienating second nature) decides.

The invitation we're doing here, then, echoes what already characterised the *critical theory* of the Frankfurt School, namely the denunciation of the *dominion* over Nature per part if the alienated objectivity of the (in)human system. The historicised idea of Western thought (even up to Marx) has always been that nature is basically just a *resource*, to be adopted and exploited for man's needs, one thinks of the transformation of nature through labour. Whereas in critical thinking Nature has once again become an equi-dignitary *subject* to which to relate to and to be respected (especially for Adorno). The re-subjectivation of nature, actually is the re-subjectivation of man because in "Nature" there is also the *inner nature*, that is, the horizon of meaning and sense (and not only the pulsional one) of man and of the biological (even before human) universe. Such re-subjectivation, in fact, implies a redemption not only (or no longer) from the repression of instincts (Freud's well-known *uneasiness for civilisation*), but also from the dramatic *removal* of meanings and sense (according to Jung) which always leads the individual back to an "order" based on what the system can codify (technically, monetarily, morally, legally).

This has certainly brought progress, but by subjugating nature inside and outside of man, this has also produced the relapse into the barbarism of a progress-regression, insofar as it is only moved by the aforementioned Heideggerian *deliberate will*. As an expressionistic portrayal of this, we recall the watercolour of Paul Klee's *Angelus Novus* ri-semantised by Walter Benjamin in his eponymous essay in *Angel of History*[8]. Civilisation certainly cannot be thought of identifiably as *evil*, because in it both the death of nature and the enrichment of life are inextricably interwoven. And it is not just a technical enrichment, but about relationships and semantical, distancing man from the immediacy of a naive and archaic relationship with things. The dialectical vision then precisely allows a point of (meta)observation that avoids being at the mercy of what is created and thus being objectified in turn. The thought that drives contemporary positive social expression and a science collapsed into *techne* is, however, a *de facto* totalitarian thought on the side of alienation insofar as it is

---

[8] *Angelus Novus*, Walter Benjamin, italian translation by R. Solmi, Einaudi, 2006.

promoted by, and only promoting, the unfolding of general (falsely transcendent) *equivalents*, of which individuals make themselves the means to their realisation. And this is why the Frankfurt School is not infrequently regarded by contemporary intellectuals as the 'enemy of science' and thus of progress[9]. This misunderstanding of the preciousness of what has been matured by thinkers such as Adorno, Horkheimer, Marcuse, Fromm, Benjamin and others is one of the most terrible sabotages to the *chances* that scientific thought could benefit to acquire a consciousness superior to the mere unfolding of *power* without any mediation/meditation, and is one of the most significant expressions of the *pseudo-formation* (*Halbbildung*[10], with Adorno) that characterises contemporary culture and many of its conformed exponents.

It's undeniable, the approach that drives current scientific research is well expressed – unfortunately – by a famous phrase by H. James Harrington: «If you cannot measure something, you cannot understand it. If you cannot understand it, you cannot control it. If you can't control it, you can't improve it»[11]. No one argues that this is wrong or useless, yet Musil's example of Galilei makes understand how just, or even, the calculation and measurement of the fall of a body say nothing about *why* it happens but describe 'only' *how* it happens. Thus, if such a view of things may be considered adequate for the kinematics of an aircraft, the motion of electrons in a semiconductor crystal, or the kind of plastic deformation of a metal alloy under mechanical exercise, it becomes at least 'problematic' when we adopt such a stance in biology and medicine, or more generally, in the relationship with Nature.

To think that one can arrive at the *knowledge* of Nature, and therefore of biology, and that can *understand* it through measurement is a first major problem. To think that, once this is done, one can *control* it is a second big problem. To think of *improving* it is the seal of a madness that has obtruded and demolished the awareness that "improving" always implies an 'oriented' process and thus a judgement based on the privileging of limited aspects that, when 'improved', will necessarily lead to the depletion of 'others'. But biology does not work this way: that is, it does not

---

[9] See (*The new animism of philosophy versus science*) *Il nuovo Animismo della filosofia contro la scienza*, Carlo Augusto Viano, pp. 94-103 in MicroMega Almanac of Science, 5/2015.

[10] *Theorie der Halbbildung*, Theodor W. Adorno, Suhrkamp Verlag Frankfurt am Main, 1972; in Italian: *Teoria della Halbbildung*, from the series (*Philosophy of Buildung*) *Filosofia della formazione / 7*, directed by Mario Gennari, *Il Melangolo* publisher, Genova, 2010.

[11] *Business Process Improvement: The Breakthrough Strategy for Total Quality, Productivity, and Competitiveness*, H. James Harrington, McGraw-Hill Professional Editore, 1991. Quotation reported at the web link: https://www.goodreads.com/author/quotes/42617.H_James_Harrington.

work 'by parts'. The evolution of a given set of organisms (e.g.,, a "species") neither takes place on certain selectable degrees of freedom (e.g.,, length of a limb, muscle mass, speed, resistance to a toxin, etc.) nor it concerns only those organisms on which we are concerned. The evolution of biological systems **is** the evolution of ecosystems: it's not a species that evolves, but always a network, and everything moves holonomically. This *modus* remains unfeasible as long as the actor (and author) of the (now synthetic) "project", i.e., human being, experiences himself as "a part" and operates on N parts at a time. And there is no other way to do this as long as one measures the world and acts through causal, particular and intentional processes such as all those that inhabit the praxis of *experimentum* and *technology*.

Then, even while attempting to investigate it and resolve its ever more and more intimate details, it would be a good idea to approach Nature with an attitude that is always *sacredly contemplative*, in the words of R. Panikkar:

> «Contemplation is something definitive. It is something that is connected with the very purpose of life and is not a means to something else. The contemplative act has in itself its own raison d'être, its own foundation. Hence it cannot be manipulated to achieve another end. In that sense it is not a stage. It has no ulterior intentionality. It requires innocence, since the very will to achieve contemplation becomes itself an obstacle to its attainment and results in its direct negation. Such an attitude is in stark opposition to the orientation of modern (Western) civilisation, both religious and secular...»[12].

Otherwise, it would mean that peoples like the Aborigines of Australia, the Pygmies of Africa, the Navajos in North America, the Yanomami of the Amazon have never understood Nature, since they have never 'measured' anything by it. It is interesting to note, as a matter of fact, how the only disasters on a planetary scale have only been carried out by that (technological) 'people' who, a few centuries ago, began to do so and who, it seems, still do not know Nature.

Today, we have *synthetic biology*. One of the most 'authoritative' science journals, *Nature* (ironic, at this point, its name) defines it as follows: «synthetic biology consists of the design and construction of new

---

12 *(The New Innocence) La nuova innocenza*, Raimon Panikkar, Quaderni di ricerca Collana a cura del Centro di Studi Ecumenici Giovanni, p.77.XXIII, pp. 77-101, Priorato di S. Egidio - Sotto il Monte - Bergamo, CENS, Cernusco s/N (MI), 1993.

biological parts, devices and systems, and the redesigning of existing, natural biological systems to make them useful for some purpose»[13]. All this is accompanied by great enthusiasm for the fact that this vibrant branch of science will radically change the world we live in. This, unfortunately, is sure.

But, besides wondering who and why would want to make this world some way... and what that way is (increasingly synthetic?), there are many questions to be asked. First, the question about what purposes could be pursued that are not already included (and therefore fulfilled) in that *choral relationship* that sees the species all contextually consonant with the (already choir-like teleological) *Life-system*. Another question concerns how and why 'purposes' heteronomous to that biological relationship should have arisen, appearing within a species called *Homo Sapiens Sapiens* that is necessarily innervated in that Life-Nature. In the following, we will offer some reflections that will attempt to shed light on this.

If by "Life" we mean that holonomic phenomenon, intrinsically collective and plural, taking place within Nature (within *physis*), we must immediately state that the distinction between "natural", understood here as "logical-bio" (i.e., in accordance with, sensible for, Life), and "artificial" goes beyond the simplistic, and cynical, position according to which "everything that exists is 'natural' because it is in Nature". According to this position, all actions and wills should be indiscriminately accepted as they are, because they are part of the evolution of the cosmos. Yet, the distinction between natural (bio-logical) and artificial, as defined above, finds profound justification in relation to the context considered, i.e., with respect to the scale, the space-time range that it makes sense to consider in order to speak of living dynamics, of ecosystems, of Life on a planet, and not only of Nature (understood instead as *physis*, reality, what that exists). Here we are talking about planet Earth, and we are talking about the living systems that inhabit it. And we are not talking about the entire universe, as a macrosystem in which everything that happens is certainly 'natural', because it is within Nature, but not necessarily 'bio-logical'.

For instance, the (accidental or intentional) introduction of living beings, developed along the choral phylogeny in a given context, into another one that is alien to them, may be a natural event, in the sense that it happens in physical reality, but it is not necessarily 'bio-logical' (whether they are ecosystems within the same planet, or, even more so, whether they belong to distinct planets). This is true because such sudden variation

---

[13] http://www.nature.com/subjects/synthetic-biology.

– such as the arrival of a *foreigner* species – produces states and modifications to which a living collectivity, or an ecosystem, may not be able to preserve itself (alive), or may be able to do so at immense costs for biodiversity and complexity (two inescapable cornerstones for talking about *living dynamics*).

Or, another example, even a cataclysm, which in extreme cases can destroy an entire planet with all the life that has developed (even with efforts) on it, is a 'natural' event. But beware: it is not a phenomenon of Life, but rather something that can happen to Life and towards which Life will try to adapt as best it can, always seeking balances that allow it in any way not to disappear.

We are aware that this attribution of subjectivity and teleology to Life may seem naively vitalistic or at least arbitrary. But if we consider this phenomenon well, Life, even if considered as a special state of condensed matter (the being in a *living phase* of matter, as we will mention later), is not a condition that can be arbitrarily attributed to numerically minimal portions of components (material and energetic): it is intrinsically choral, it requires a minimum degree of multiplicity and numerosity. For example, we cannot have a living cell made up of a few hundred molecules, nor of a few thousand. Or, similarly, a life form, a species, does not live if it does not share its context of existence at least with some others, and it is well known that an ecosystem is only stable if a minimum number of (variable) species alternate and interact in a given region of spacetime (both on a cellular and macroscopic scale).

Thus, Life is not a phenomenon *of* the living or *of* the species but unfolds *through* (*by means of* and thanks to) them, plurally. From the totality of the living systems, it is constituted and at the same it time transcends them, since it is, if you like, their *interrelation*, it is their *dynamic* and (we could say) *autopoietic equilibrium*, since it tends to maintain itself and restore itself (although without identity to previous states). That "something", without which, their very being alive (their *beingness*) ceases. In this equilibrium, the aim is never the prevalence of one species over the others or the possibility for their plurality to be compromised since this would mean destroying the interrelationship itself and weakening Life as a choral phenomenon.

Therefore, when we speak of *biological defaults* in our biosphere (or in a more naïve way, of modes/gestures "according to nature"), we are referring to what that it's possible to do within the living realm and, therefore, what makes sense to maintain the balances that allow Life to perpetuate itself, meant as the *collective phenomenon* of existence, of 'being there', proper to living beings and to their dynamic becoming and changing on this planet. And indeed, the tendency to maintain this 'beingness' is the only true fundamental characteristic and, if you like, the only true universal *teleology* and "sense" of Life (plausibly everywhere in the universe).

With regard, therefore, to the technological gestures performed by human beings within and on the realm in which Life takes place (i.e.,, acted upon both directly on living beings – humans themselves or other species – and indirectly on contexts, resources, spaces, etc.), we can discern between acceptable (i.e.,, still "bio-logical") and unacceptable (or "artificial", "unnatural") modalities in relation to:

i. the extent/scale of the effects they have on the realm of terrestrial Life and its sub-ecosystems and its spacetime context (nature, in the broadest sense), even in the long term and over a long range, and

ii. the *aesthetical* (perceptual) stance with which such technological gestures are conceived and performed, i.e.,, whether or not they originate from a sensibility that could be described as "indigenous-like". Let us explain ourselves better.

In the case where gestures are acted out in a 'native-like' position, these are actions mediated by a sensibility that is based on *listening* and the recognition of a *deep intelligence* in the balances that Life has evolved. This is what has typified, for example, the behaviour of those human societies in which the relationship with nature is, we might say, of *"neolithic mould"*, as it was for the aborigines, the pygmies, and for certain last tribal cells in Amazonia or Papua New Guinea.

Therefore, even in the case of actions aimed at 'defending themselves' from adverse circumstances or aimed at modifying the environment or producing tools for the realisation of certain purposes, that sensibility has always given those forms of human civilisation a healthy *demeanour* such as to express their actions with 'reverence/caution' to a greater balance and in respect of a *purpose* felt to be superior to the "personal wishes" of humans (i.e., the maintenance of Life, with its own balance). And this was possible because the indispensability of this balance for one's own existence and that of the ecosystem has already been recognised and, therefore, its profound *sacredness*, all biological (and not so much mystical), has already been grasped.

To give another example in which we can grasp the difference between being and not being endowed with this sensibility-demeanour-reverence, let's think of the bio-logical option of building bows, arrows and spears to hunt, say, bison across the North American prairies (as the American natives did), as opposed to (non-biological) equipping hunters by automatic rifles used by the colonists, which suddenly allowed the extermination of half a cattle in a single hunting trip. And if we want, the substantial difference lies not only in the technical means (the weapon used), but in the gestural expressiveness by which they are used: the natives, even when equipped with rifles, never practised that way of hunting because they were already educated, connected, to a sacred relationship with the womb of Life. The latter, as we well know, have impoverished natural environments countless times and in various epochs.

In this writing, we are making a heartfelt appeal to reconnect with a profound *centre* that inhabits every human being (and not only) beyond any culture, language, territory, epoch, and which is the only *perceptive core* that can strip everyone of all ethical, religious, cultural, juridical, economic, and psychological conditionings, constituting the artificial identifications from which the world is viewed, in order to feel the *true-good-beautiful* quality intrinsic in the *logos* of Life and respect it. Recapturing that perceptive kernel allows us to dis-identify ourselves from the superstructural cultural, ideological, narcissistic constructs and emotional imbalances, leaving aside the deliberate 'I want' that increasingly dwell in the inauthenticity of the deformed culture of capital, technology, 'rights', 'law' and all the *fictional superstructures* which alienated the *homo technicus*.

These superstructures (which disguise that biological sensibility that is being invoked) have always been believed to be necessary to produce an appropriate social order, freeing men from a postulated destructive ferocity. And their collective, positivised correspondents (such as state, law, money, market, science, religion, etc.) are nurtured because they are believed to be the only guarantees of social living, because they are believed to be the only solutions to the problems caused by a (Hobbesian) 'nature of man' that, left to its own, would lead to catastrophe, to bestiality.

In truth, such structures (both structural of the social realm and psychic of the subjective one) are thought of as the solutions to those problems of which they are the first essence, since they are the bearers of an 'original sin' given by the non-reconciliation with that natural *connectome* mentioned above, by the loss of that sensitivity. Thus today's 'civilised' humanity is behaving like someone trying to fill a tub of water to the top by only increasing the inflow and not closing the drain plug. We will return to these themes in various chapters and in the conclusions.

As we were saying, gestures that are not in the name of that care, caution and reverence are not "good" and, note, this *goodness* **is not ethical** or cultural, nor ideological since it is not related to when, to where, nor to values heteronomous to Life itself, in its most bare and essential meaning. It is a *goodness* that is not based on any opinion, nor on who explains it. **It is all biological** because it pertains to being in tune with the intrinsic *qualitas* of any living ecosystem (even planetary) about self-tending for its own (dynamic and ever-evolving) maintenance, to sustain Life. "Life" which is not "of" someone or someone else, of certain species or others, but rather 'Life' which is in fact that dialectical correlation between all species (and their environments, niches, evolutionary trajectories) and which constitutes the *beingness* of organisms, albeit in the alternation of disparate balances that see the extinction, appearance, transformation and merging of the previous ones. In a word, the choral and holonomic evolution.

All this, in the biological-natural *default*, is done without a biased position or perspective, that is, from someone's point of view. It does not have the bias of a *decontextualised will* (as the 'willing' human being may have in pursuit of a goal). And thus, it has no heteronomous purpose to that *logos* of Life that Life itself 'wants' to preserve. This is what we mean by 'natural'.

When the action run within and on the realm of Life (indirect or direct) is instead "intentional" from a "point of view" (for instance of a single species, such as man) that pursues aims conceived in an aborted, conditioned sensibility, oblivious of respect for this "over-determined syntax", for this greater balance, without this care, then such action is said to be "artificial" in negative sense. Each time such a position must be contextualised, clearly, and it is precisely that sensibility, that care, that listening, which allows one to always orient oneself in tune with Life. No modality can be crystallised once and for all or enshrined in any rule or protocol, it must be recaptured as a *breath* that is already here. In this sense, it is necessary for the natural sciences (from physics to medicine) to seriously come to terms with the analogical feature of nature as the constituent essentiality of the living realm.

In four and more billion years of phylogeny, the human being is the first species that has the (technical) power to potentially put an end to all Life on the planet or to manipulate it at its own faculty and will; and it is the first species that manifests the intentional aptitude to pursue aims heteronomous to the very essence of the systems on which it operates without an endogenous demeanour (if not moral or religious) of preserving the environment and 'that balance of all Life' on which the human being himself depends (planet Earth). It does not seem too bizarre to contemplate the hypothesis that human being is in part 'foreign', that is, belonging to a different biological paradigm, proper to *elsewhere*[14].

---

[14] In this regard, we suggest an open-minded and unprejudiced reading of ("*Rendered Humans*") *Resi Umani*, by Mauro Biglino and Pietro Buffa, Uno Editori 2018.

1. What is happening?

"Here is a new semi-synthetic life form" and *"When the code of life is written with six letters"* spoke the titles of the articles on the 25th of January 2017 post[15] by *Wired.it* (web magazine) and the 27th of January 2017 newsletter[16] by *Le Scienze*, respectively. The reference, in both communications, is to the work published[17] on PNAS the (*Proceedings of the National Academy of Sciences of the United States of America*) on behalf of a group consisting of five researchers from the Scripps Research Institute (California), one from the Université Grenoble Alpes (France) and one from the Henan Normal University (China). First signature and coordinator of the project is Floyd E. Romesberg, from the Scripps Research Institute, a non-profit, private research facility specialised in biomedical sciences.

The topic is about the successful synthesis (already achieved in 2014) and stabilisation of *Escherichia coli* bacteria (or perhaps one could not even call them anymore like that) whose genome was composed of six bases: i.e., the four "natural" ones (adenine, thymine, guanine, cytosine) plus two synthetic ones (called X and Y given by the nucleotides 'd5SICS' and 'dNaM') inserted *ad hoc* via bio-engineering techniques including the recently developed CRISPR-Cas system.

The approach that animates the working group and the 'consideration' by which it relates to the object of study (in this case: biological systems and, therefore, Nature in its emerging aspects) is, in our opinion, emblematically expressed by the first line of the *pre-abstract* (section "Significance") of the PNAS article: «The genetic alphabet encodes all biological information, but it is limited to four letters that form two base pairs». Particularly interesting is that «*but* it is limited to four letters» [italics of ours], referring to the genetic "alphabet". That same alphabet by which biological phylogeny has been articulated for about 4.3 billion years and thanks to which it has been able to express itself into a multitude of living species and, above all, living *systems*, ecosystems, always

---

[15] https://www.wired.it/scienza/lab/2017/01/25/nuova-forma-vita-semi-sintetica/

[16] http://www.lescienze.it/news/2017/01/24/news/ingegnerizzare_organismi_semisintetici-3394075/?ref=nl-Le-Scienze_27-01-2017 ;

[17] *A semisynthetic organism engineered for the stable expansion of the genetic alphabet*, Yorke Zhang, Brian M. Lamb, Aaron W. Feldman, Anne Xiaozhou Zhou, Thomas Lavergne, Lingjun Li, and Floyd E. Romesberg, vol. 114 no. 6, 17–1322, doi: 10.1073/pnas.1616443114, http://www.pnas.org/content/early/2017/01/17/1616443114.

endowed with a profound intra- and inter-direct correlation. We speak, therefore, of that 'molecular quatern' – now defined as "limited" in the eyes of *Homo Technicus* – which has allowed "Nature" to move always holonomically and with extraordinary biological consistency, showing us that what exists are *processes*, rather than living *entities*, and that what evolves are not *species* (which, moreover, are able to share large genome sequences[18]), but *systems* and their *relationships*. It is not only the neck of the giraffe's ancestor that elongates, but there is an ecosystem that evolves its configuration (we might say through phase correlations[19]) and, as a result (among many other aspects, more subtle to grasp) there is also the elongation of the neck of a species that we will then call 'giraffe'. It is a pity that these actualities cannot be 'measured' except when they bring about large changes and on (at least) the 'human-scale' space-time scales.

And while these same text-lines unfold on the white pages, the Chinese studies on the already created (and alive) calves genetically modified[20] to be "tuberculosis-resistant" and the, at least terrifying, pre-publication – published by the National Academies Press – of the guidelines for the modification[21] of the human genome… burst into the scientific literature. And in the catalogue of genomic games, we can also mention the synthesis of 5 synthetic chromosomes in yeast[22], the creation of an artificial mouse embryo[23], the bacterium regulated by a synthetic genome[24]. And who knows how many more "scientific victories" will see the light between now and when someone reads these pages. Obviously, even these "novelties" are peacefully reported, by those who popularise and make the public aware of what is moving on the scientific research front, without any critical attitude about the gnoseological premises that make these

---

[18] *Gene Transfer Between Species Is Surprisingly Common*. University of California - Berkeley. ScienceDaily, 11 March 2007. www.sciencedaily.com/releases/2007/03/070308220454.htm

[19] Brizhik Larissa, Del Giudice Emilio, Jørgensen Sven E., Marchettini Nadia, Tiezzi Enzo, op. cit. *Coherent Quantum Electrodynamics in Living Matter;* Emilio Del Giudice, Antonella De Ninno, Martin Fleischmann, Giuliano Mengoli, Marziale Milani, Getullio Talpo, and Giuseppe Vitiello, Electromagnetic Biology and Medicine, Vol. 24, No. 3: Pages 199-210, 2005; *The interplay of biomolecules and water at the origin of the active behavior of living organisms;* E Del Giudice, P. Stefanini, A. Tedeschi, G. Vitiello; Journal of Physics: Conference Series 329 (2011) 012001.

[20] *TALE nickase-mediated SP110 knockin endows cattle with increased resistance to tuberculosis,* Haibo Wu, Yongsheng Wang, Yan Zhang, Mingqi Yang, Jiaxing Lv, Jun Liu, and Yong Zhang, vol. 112 no. 13 E1530–E1539, doi: 10.1073/pnas.1421587112, http://www.pnas.org/content/112/13/E1530

[21] *Human Genome Editing – Science, Ethics, Governance,* National Academy of Science – National Academy of Medicine, https://www.nap.edu/catalog/24623/human-genome-editing-science-ethics-and-governance.

[22] http://science.sciencemag.org/content/355/6329/1040/tab-pdf

[23] http://science.sciencemag.org/content/early/2017/03/01/science.aal1810.full

[24] http://science.sciencemag.org/content/sci/329/5987/52.full.pdf

results appear as "steps forward" in research or as solutions to the problems of humanity and the planet.

As E. Severino has already reminded us, «one only wants to discuss immediate problems and one only recognises a meaning to problems if one can already glimpse the specific techniques for solving them»[25]. Therefore, they only want to deal with solutions placed on the same plane on which those same problems, that one believes to solve, lie. It happens right what Paul Watzlawick calls *type 1 change*, and thus «plus ça change et plus c'est la même chose»[26].

The *solutions* to be undertaken, then, concern the one and only real problem, which, however, is not taken into account at the level of those who look for solutions (or tell about doing it), namely: the fundamental and implicit *modus* of approaching the world, one's own reality, Nature and, consequently, the investigation of it, hence *how* Science is done, which underlies the worldwide mainstream of scientific activities and the entire 'profane' population regarding the total lack of a "right fit" in the scientific *praxis*. A measure, as the Latins already taught us (*est modus in rebus*[27]), which must be adopted for every gesture of existence, in theory as much as in practice, and especially when it comes to biological *logos*. That from which we cannot prescind. Never.

To consider as a "limit" the fact that there are four (rather than n>4) gene bases, evolved by Nature's self-consistently in that precise manner, or to consider as a "limit" the fact that our species (or any other) has some faculties whose reach extends up to a "point" (naturally gauged) and not up to another one, idealistically (and ideologically) placed, means having understood a damn nothing about the world. It also means having believed in (and succumbed to) the myth of the isolable entity, of the part in itself, it means having forgotten all that is important for life and for its true enjoyment, which is only possible insofar we're *connected* to a sophisticated and inescapable realm of relationships.

Within this oblivion many attitudes are included which have as their idol the efficiency, in its disparate forms, often dressed up as (and/or hybridised with) the protection of 'rights' or of interests to which the removal of what is actually happening comes along, i.e.,: the unconditioned and deliberate self-imposition of *pure will* oriented only towards itself and

---

[25] See (*The Essence of Nichilism*) *L'Essenza del Nichilismo*, Emanuele Severino, *pag. 263*, Adelphi, 1972.

[26] See (*Change. About the Formation and the Solution of problems*) *Change. Sulla Formazione e La Soluzione dei problemi*, P. Watzlavick, John H. Weakland, Richard Fisch, Astrolabio, Roma, 1974.

[27] Horace, *Satires* I, 1, vv. 106-107.

its own fulfilment (see the incipit quoted by M. Heidegger). Let us take a few examples: why to subsume that everyone in a species must necessarily have progeny? Why not try to adopt a species-consciousness that knows how to appeal to a healthy, ethical, mature, self-listening and self-acceptance and recognition of biological intelligence that is always consonant with the *boundary conditions* and to love ourselves independently of roles, ideas about being "this or that" and conditioning projections? By what criteria do we decide that certain faculties of the human brain or organism should be enhanced? And what does 'enhanced' mean? And what consequences would this have on the entire organism and its relations with the world? Haven't we yet had enough of how much "out of the choir" our characteristics are in relation to the other species in the biosphere, and of how much our careless and presumptuous expression has already made us responsible/guilty for countless disruptions? Why not to embrace, with a little humility, what Nature knows for self-regulation?

Changes in every species, according to natural developing, occur through progressive (and collective!) adaptations and evolutions[28]. How can we put the hands on something believed to be manageable through a very limited 'molecular vision' that could be considered advanced no later[29] than the 1950's? Why always to reduce every expression of a (supposedly ontically detachable) *hardware*-soma only to a computable *software*-genome (limitedly supposed digitizable, material and now also upsettable, reprogrammable)? Do we want the (almost Orwellian) myth of the *total cybernetics* to overwhelm our every faculty of *visioning* reality? Do we really believe that *biosis* unfolds and originates in accordance with the 'biochemical version' of the functioning of DNA (as is taken for granted in mainstream biology and genetics), neglecting all its electromagnetic, morphogenic, biophotonic (and who knows what else) roles? In what one does not have complete knowledge and mastery of, one should not intervene. Especially if the consequences are imponderable in their extent and/or radicality.

---

[28] *The role of electromagnetic potentials in the evolutionary dynamics of ecosystems*; Larissa Brizhik , Emilio Del Giudice, Sven E. Jørgensen, Nadia Marchettini, Enzo Tiezzi; *Ecological Modelling 220 (2009) 1865–1869*; *Morphic Resonance*, Rupert Sheldrake, Inner Traditions Editore, 2009; *The Science Delusion*, Rupert Shaldrake, Coronet Editore, 2013.
[29] *Bioenergetics*; A. Szent-Gyorgyi, Academic Press Inc.: New York, NY, USA, 1957

2. Synthetic spells

Note well that we are talking about the same problems as GMOs, artificial intelligence, babies made in vials, wombs for rent, abortion and, like it or not, animal experimentation and euthanasia. And these problems are not of a technical order, they do not concern the procedures to be followed in order to be successful in the laboratory, *in vivo* or *ex vivo* (however ambiguous what "successful" means now, in a large-scale view). And these are nor ethical-moral or legal issues, those depend on and arise from heteronomous cultural constructs that do not exist in themselves but just *value*; and may apply here, but not there, and yesterday, today, but not tomorrow.

While to the previous aspects it must also be given the due consideration, first we must deal with a problem of far more fundamental necessity. This is about that sensibility and care that does not pertain to any form of culture or thought or religious belief or philosophical adherence: we are talking about the ontological *biology*. That is to say, of the biological *logos*, of the biological *speech/syntax*, of how the *bios* is unfolded and carried on in physical reality, of the fact that we are dependent on it, that we are immersed in it, and of the fact that if we place ourselves 'above' that "speech", altering its syntax to our own idea, we commit an irreparable crime because, on that terrain, there is no way back from any step. Once the system (in fact a holo-system) has been perturbed, it is not possible to backtrack along an elementary linear causal chain and "fix back the play". Yes, because the correlations are virtually infinite, very subtle and very long ranged. Put in other, technical terms: the bifurcations in the phase space of a highly correlated system, in which some perturbations can act as attractors, cannot be travelled backwards, i.e.,, the trajectories in the phase space do not connect the same states regardless of the direction of travel (they are not isotropic). It is not enough to remove the perturbation once the system has entered a bifurcation. Once the red button is pressed, there is no going back.

So, taking the work published in PNAS as the main starting point, the first question to ask is unfortunately also the one that 'science workers' in particular, do not seem to ask themselves enough: do we have even a vague idea of what it means to 'give life' to an organism with a genome whose 'syntactic criteria' are completely alien to the entire context of the biosphere? It seems that those who work in bio-engineering fields do not have it quite right, and we shall see why.

Let's proceed step by step. The problem concerns all the technical-scientific expressions that denote not only the lack of clarity about the fact that the human being is not "above nature", but also a profound coarseness in the representation made of how nature functions: at the extremes we find (at potential or current stages) the creation of organisms with a genome with a different genetic alphabet, GMOs, the modification of the human genome on both somatic and germ cells. And other critical issues that will be discussed below.

There are therefore two aspects that arise in association with these circumstances, one, perhaps, more serious than the other: the first is that there are scientists, working in research centres, who have the desire, the idea, the motivation (and the concession) to spend their resources in these directions believing themselves to be in an intellectually (humanly!) correct and sustainable position; the second is that those who do popularise science (editors and journalists), report the results published on such topics without introducing due reflections on the (at least potential, in reality definitely actual) problematic nature of what they are reporting. A bit like someone who, for instance, when referring to the discovery of the quantisation of electronic states in an atom, is not also considering what implications it has, at least as a hypothesis, and does not mention it at all.

The only "remark" that is made about the goodness or not of this type of bio-technological practice appears in the article[30] by *Wired.it* only as a "chronicle of the facts" and not as a *thoughtful* reflection on the subject of the article: it mentions that «there is no shortage of concerns about the potential impact which this type of technology could have: back in 2014, for example, Jim Thomas of the Etc Group, a Canadian organisation that analyses the socio-economic and ecological impacts of new technologies, had explained to the *New York Times* that *"the arrival of a semi-synthetic life form could over time have unpredictable repercussions on ecosystems, not to mention the ethical and legal aspects of the issue"*».

Romesberg's retort to this initial criticism of the Etc Group (which, in our opinion, was however not incisive enough) – it is reported – is on the one hand about the fact that we are (were!) still a long way from implementing such a modification in organisms more complex than *E. coli*, and on the other hand about the fact that the work carried out by his research group is basically nothing more than a replication of what happens

---

naturally with evolutionary processes; here is his final quote: «Evolution, [concludes Romesberg], works by modifying existing organisms in small steps. Just what we have done" ». (What?!)

Now, if it were not evident enough in itself, these are unreasonable and dangerous statements (!): how can it be claimed that such bioengineering is the equivalent of retracing the biological evolutionary process in small steps?! Would introducing two new nucleotides into the genome and changing the protein syntax from a 4-element code to a 6-element code be a 'small step'?! Secondly, to say that we are still a long way from implementing the new bases in the genomes of more complex organisms is doubly serious: on the one hand, it does not mean that this practice does not already constitute a problem, given that the possible damage (to which we will refer below) is not proportional to the complexity of the synthetically produced organism, but to the multiplicity of biological correlations in which that organism may participate; on the other, it seals, without a shadow of a doubt, that the direction being aimed at is in fact right that of extending such genomic 'implementation' to other species (!). Here is what is reported a few lines earlier: «Romesberg, for example, has imagined a future in which insects are capable of producing medically useful proteins, similar to therapeutic insulin. *The synthetic DNA we have developed and inserted into bacteria,*" the author explains, "*could one day help us to create organisms that possess entirely new and unexpected characteristics and abilities. In this sense, our research represents a big step towards our long-term goal of creating semi-synthetic life*" ». We repeat, in case it was not clear enough: «... a big step towards our **long-term goal**, namely **the creation of semi-synthetic life**'» [emphasis added].

3. Seductive and senseless aims, rapid and uncontrollable means

Our ignorance of social realities comes from our tendency to focus solely on modern societies. We rarely consider the idea that social experience in the Palaeolithic era can teach us anything, even though it probably constitutes 90% of the human experience on this planet! To think that we are dispensed from the laws that govern it by relying on science and technology is an act of faith that serves to justify the systematic violation of those laws by the industrial activities to which our society is subjected.

Edward Goldsmith, *L'Ecologiste,* 27, Autumn 2008, Vol.9-3, p.15

Now, the risk that the most apprehensive or the most opportunistic (such as the long-eyed businessmen) would greet such a novelty with enthusiasm is high. The apprehensive would do so because of the hoped-for comfort of greater health guarantees; the opportunistic because of the seductive prospect that an active ingredient, which today costs perhaps 10,000 dollars per gram (possibly due to catalytic acrobatics, spurious products, and enantiomers to be discarded), could, for example, be synthesised by semi-synthetic bacteria, cutting costs by two or three orders of magnitude.

And what's wrong with hoping to have more guarantees and/or "tools" on the healthcare front, or in aiming to drastically cut costs to produce molecules useful for human activities? Certainly, these two questions, as posed this way in themselves, can only lead to the answer "there's nothing wrong, nay the opposite!" But we must see both (i) whether the end justifies the means, (ii) and whether the end has a real and inescapable necessity. Let us explain further.

The argument about the increased pharmacological-medical chances is flawed by the presence of an error of method in medicine – both the self-told 'scientific' medicine and much of the so-called 'complementary' medicine – which, for the same reasons for which 'science events' such as those we are reporting here occur, is unable to promise significant breakthroughs in its own performance and efficacy since it is participant, and consequence, of an epistemological and technical context that has not yet come to terms with *what* 'disease' is, since it has not yet begun to consider living systems as semantic processes *e fundamento*. This refers to the prejudicial renunciation, for methodological (and ideological) reasons, to recognise the real causes of 'pathological' processes down into the neurovegetative dynamics of *semantic interfacing* of a living system with its environment. Whether or not it is more or less aligned with the mainstream, this pertains to what has already been investigated or traced with great acumen by scholars such as H. Selye[31], W. Cannon[32], J. Mason[33], H. Laborit[34], S. Porges[35], B. Lipton[36], C. Pert[37], not to mention the incredible work on the causal connection between biological (not psychological!) perception and programmes in special physiology conducted by R.G. Hamer[38], condensed into the 5 biological laws he finally acknowledged.

In spite of the sceptical (and cynical) attitude of many who are content with a view of science firmly entrenched in epistemological reductionism and of those easily swayed by what echoes in shoddy, when not hypocritical, journalism, the picture, offered by these studies and others, expressing what is the functional basis of biological systems, is anything but naïve and is well supported by (i) the description of the latters within the consistent categories of a field approach (as in Quantum Field

---

[31] *A syndrome produced by di-verse nocuous agent,* Hans Selye, in Nature, CXXXVIII, n° 32 (1936).

[32] *Bodily Changes in Pain, Hunger, Fear and Rage,* vol. 2, Cannon W., Appleton, New York (1929). *The Wisdom of the Body,* Cannon W., Norton, New York, (1932), Tr. it. *La saggezza del corpo,* Bompiani, 1956

[33] *A re-evaluation of the concept of "non-specificity" in stress theory,* Mason J.W. in Journal of Psychiatric Research, VIII, n° 323 (1971).

[34] (*In Praise of Escape*) *Elogio della fuga,* Laborit H., Mondadori, Milano (1990).

[35] *The Polyvagal Theory: Phylogenetic substrates of a Social Nervous System,* S.W. Porges, International Journal of Psychophysiology 42 (2001). 123-46, Elsevier, http://www.boazfeld-man.com/EN/Links_files/PORGES%20Polyvagal_Theory.pdf; *Social Engagement and Attachment: a phylogenetic Perspective,* S.W. Porges, Annals of New York Academies of Sciences, 1008: 31-47 (2003), http://www.psy.miami.edu/faculty/dmessinger/c_c/rsrcs/rdgs/attach/Porges_socialEngagementAttachment.pdf, *A Review of "The Polyvagal Theory: Neurophysiological Foundations of Emotions, Attachment, Communication, and Self-Regulation,* S.W. Porges, Journal of Couple & Relationship Therapy: Innovations in Clinical and Educational Interventions Volume 11, Issue 4, 2012, http://www.tandfonline.com/doi/abs/10.1080/15332691.2012.718976.

[36] *The Biology of Belief,* B. H. Lipton, Penguin Books India, 2005, *La biologia delle credenze,* B.H. Lipton, Macro edizioni, 2006

[37] *Molecules of Emotions,* Candace B. Pert, Scribner Editore, 1997, *Molecole di Emozioni,* Traduzione L. Perria, TEA Editore, 2005

[38] *Vermächtnis einer Neuen Medizin,* Hamer R.G. (1999) Amici di Dirk, Alhaurin El Grande, Tr. it. *Testamento per una Nuova Medicina (Will for a New Medicine),* Amici di Dirk, Alhaurin El Grande, 2003. *Introduzione alla Nuova Medicina (introduction to New Medcine),* Hamer R.G. (2002), Amici di Dirk, Alhaurin El Grande; *Krebs und alle sogenannten Krankheiten,* Hamer R.G. (2004), Amici di Dirk, Alhaurin El Grande, Tr. it. *Il Cancro e tutte le cosiddette malattie* (*Cancer and all so-called diseases*), Amici di Dirk, Alhaurin El Grande, 2005.

Theory[39]) and by (ii) the acknowledgement of the crucial role of water in a super-coherent condensed state in every biochemical process[40].

The phenomena discussed and reported in the above researches are proof that the chemical reaction pathways deciding which is the physiological state in a tissue or in the whole system (normal or symptomatic) are dependent on the system's phase configurations, i.e., on the type of resonance established (in the sense of which degrees of freedom are involved and on which, time-varying, frequencies are performed) between its own oscillators and those of the context in which a living system is inserted (i.e., environment, but with both thermodynamic and semantic meaning!). This finally makes it evident that the (neuro)biological act of perception has a notably physical foundation and is centred on an observable of wave processes (the phase) conjugated to amplitude[41], the only measurable one[42]. This makes impossible any direct determination of the phase, but physics is well aware of this and biology and medicine can no longer ignore it. In extreme synthesis, which is necessary here, all of this expresses how "pathological" processes are *sensible responses* in physiology (sometimes mediated by other more or less symbiont organisms) activated by the autonomous neurovegetative system, whose primary purpose is the survival of the living being[43]. Suffice it to say that denervated

---

[39] *Advanced Field Theory: Micro, Macro and the Thermal Concepts*, Umezawa, H. American Institute of Physics: New York, NY, USA, 1993; *My Double Unveiled*. Vitiello, G., John Benjamins, Amsterdam, 2001. *My Double Unveiled*. Vitiello, G., John Benjamins, Amsterdam, 2001; *Quantum Field Theory and its Macroscopic Manifestations*, Blasone, M., Jizba M., Vitiello G. Imperial College Press, 2011

[40] See references in footnote 19 and: *QED Coherence in Matter*, G. Preparata, World Scientific Publishing Co, 1995; *A Quantum Field Theoretical Approach to the Collectiv Behaviour of Biological Systems*; E. Del Giudice, S. Doglia, M. Milani, Nuclear Physics B251 [FS13] 375-400 (1985); *Coherent Quantum Electrodynamics in Living Matter*; Emilio Del Giudice, Antonella De Ninno, Martin Fleischmann, Giuliano Mengoli, Marziale Milani, Getullio Talpo, and Giuseppe Vitiello, Electromagnetic Biology and Medicine, Vol. 24, No. 3: Pages 199-210, 2005; *Water and the autocatalysis in living matter*, Del Giudice, E.; Tedeschi, A., *Electromagn. Biol.Med.* 2009, 28, 46-54. *Water: A medium where dissipative structures are produced by a coherent dynamics* Marchettini, N.; Del Giudice, E.; Voeikov, V.L.; Tiezzi, E., *J. Theo. Bio.* 2010, *265*, 511-516. *Coherent structures in liquid water close to hydrophilic surfaces*, Del Giudice, Vitiello, Tedeschi, Veikov, Journal of Physics: Conference Series 442 (2013) 012028

[41] If N is the number of oscillators, Φ is the oscillation phase and Δ is the relative uncertainty, we have (in MKS units): $\Delta N\, \Delta\Phi \geq 2\pi$. See: *QED Coherence in Matter*, G. Preparata, World Scientific Publishing Co, Pte, Ltd, 1995.

[42] Preparata G., op. cit.

[43] *Electrodynamic Coherence as a bio-chemical and physical basis for emergence of perception, semantics and adaptation in living systems*, P. Renati, Journal of Genetic, Molecular and Cellular Biology, 24 December 2020, Enliven Archive, ISSN: 2379-5700, http://www.enlivenarchive.org/articles/electrodynamic-coherence-as-a-biochemical-and-physical-basis-for-emergence-of-perception-semantics-and-adaptation-in-living-system.pdf;

tissue does not become inflamed. Thus: what we call 'inflammation' is not an 'accident' of the system, but rather its ordered and sensible response to an event (chemical, mechanical, radiative, semantic)[44,45].

It is precisely this *epistemic subtraction* of the physical dynamics underlying semantics in living beings that makes precisely that which is passed off as verifiable, reproducible and repeatable, an irretrievable and incomplete pseudo-science. The same one that, consequently, deceives us about the idea that problems can be solved within a theoretical framework that "believes in disease" as if it were an "entity in itself" that "attacks" the living system.

Here it is the way we come to design genetically modified calves 'resistant' against tuberculosis, as if we did not know (because perhaps we do not want to know) that mycobacteria are part of the symbiotic microflora of mammals. The genome change on calves does nothing more than momentarily interrupt the synergy between lung tissue and bacteria, which in those tissues have to perform functional tasks for the living (in the case, the calves) and which are modulated by the said 'immune system'. But if we do not realise that the tasks they perform (with which symptoms are associated) are a consequence of what the calves experience biologically (at the level of neurovegetative states), nothing will be resolved. Because, after a certain amount of time has passed, the GM calves, farmed under the same (in general, very bed) conditions and experiencing the same *biological perceptions*, will require the same physiological activations and processes that will still have to be mediated by symbionts, which in turn will have to evolve their genome in order to interface (again) with the changed context. In essence, not only does this not solve the problem, but it also triggers 'unintended' mutations in other living systems with consequences that no one can assess. And the same applies to the human genome, a practice for which editing regulations are even being written, as we have already mentioned in the introduction. But to borrow Emanuele Severino's words, «doesn't true health come about because one is able to discover the true illness?»[46].

---

[44] *Relationships and Causation in Living Matter: Reframing Some Methods in Life Sciences?*, Physical Science & Biophysics Journal, September 28, 2022, Volume 6 Issue 2, ISSN: 2641-9165, MEDWIN PUBLISHERS, DOI: 10.23880/psbj-16000217, https://medwinpublishers.com/PSBJ/relationships-and-causation-in-living-matter-reframing-some-methods-in-life-sciences.pdf;

[45] *QED coherence in condensed and living matter – Theoretical frameworks, experimental results and epistemological implications*, P. Renati, Doctoral Thesis, February 2021, Ph.D. Course of Complex Systems for Physical Sciences, Life and Social-Economics Sciences, Department of Physics and Astronomy, University of Catania (Italy).

[46] Severino E., *op. cit.*

All this *technological action* takes on the appearance of a 'Laurel & Hardy' sketch in which, in order to fix a lame table-leg, one hastily tries to prop it up with a napkin taken from the table, and in the gesture of grabbing it one drops the bottle of oil that was partially leaned on it; and so the oil is spilled over the table and to contain the spillage more napkins and tablecloth are hurriedly rolled up, also impacting against the bottles of wine and water which, rolling apart, fall to the floor breaking themselves in a thousand pieces and in order to pick up the glass on the floor, one furiously grabs the broom at the corner of the room, hitting hard the nearby abatjour, which falls and detaches itself from the electric wires, which, once fallen uncovered on the wet floor (with the water and wine that had previously fallen), short-circuit and blow out the light, someone is perhaps even electrocuted (!), etc., etc.. All very clumsy ... and potentially lethal.

For this reason, those who worry about their own and their children's health should be told that the solution to problems should be on a higher level than the problems themselves. And, above all, the solution should not bring with it other problems, especially if they have a planetary scale. As we shall see, introducing into the biospheric field a genome that has no roots in the biological phylogeny (since it has six nitrogenous bases and not four) could have devastating consequences.

While, on the other side, the motivation of economic relevance has the flaw of a necessity in truth only apparent that has nothing ontological about it, but is all an ideological construction, rewritable at any time: although this seems naïve, to say the least, no one has ordered human beings to mediate their material and exchange relations through a monetary system or any representation of 'value', barter included (and even if someone had done so, it would be a good idea for everyone to disobey such a scheme since it is far more counterproductive than profitable). It is good to remember that the monetary system is only a system of value representation and that, as such, it necessarily falls short, vanishes, at the level of raw resources (both material and human): in fact, one does not 'pay money' to the earth so that it produces oil, metals, or so that salad is born, nor does one give anything in return: It's enough have care and knowledge about how to do things right, we just need to create the conditions for them to happen. And people are paid for what they do or give only because they have to pay other people or entities for the same reasons (and sometimes even without real reasons, but 'by law', they say, as in certain taxations). This need to commensurate things with 'how much is their value' (moreover, purely relative to contexts) is instituted and enforced when endogenous empathic and cooperative attitudes are not adopted as pillars of a society/community. If it were necessary to produce indispensable molecules for the survival of 'sick' people, an economic system that was ethical and resource-based would produce what is needed, when it is needed, and give it to those need it. It is called *sharing* instead of barter/sale/buying/exchange. Madness? Utopia? Go ahead and believe it. We have already paid the price for centuries and we will continue to do so. Or, perhaps, they (many) will 'continue' to pay, because those who organise the global systems of markets and laws are not the same people who feel those consequences burning in their skins in terms of unequal distribution of wealth, technology, resources, services and of... boycott of the *beauty*, meant as subtraction from squalor and degradation (by natural right) and vow to the *grace* of the world and Nature.

Although men – hiding behind an embarrassing prudery the inhuman cruelty of this whole "economic" carousel – have by now become accustomed to the logic of quantitative *do ut des* (i.e.,: to give for receiving), and consider it ethically respectable and the only possible way to codify the distribution of resources, activities and services, to tell the truth, the vaunted equity and ethicality are instead sabotaged by default, since in a monetary system, as long as someone gets richer, it is because someone else gets poorer. This does not happen in a sharing system; that's not Nature's way (if looked at systemically). We are not talking about the old barter, but about something else...we let those who are able to *see*, by dropping the veils of numerical fiction, understand[47]. And you don't have to be Gandhi or Saint Francis to create such a *setting*. As Spinoza has already suggested, who acutely deduces altruism primarily from reciprocal *utility* (from which originates a growth of subjective existential powers that can be well condensed in his *homo homini deus*), it is enough to be soundly pragmatic and realistic: we cannot really be well if everyone is not well. So it would be sufficient and necessary: 1) to recognise what is needed to produce, 2) if and why (i.e., according to what idea of humanity

---

[47] It is not possible here to develop such a vast theme as that of an a-monetary economy, based on the sharing of resources and of a right of social existence not subordinated to quantified labour, but aimed at the free and vocational expression of human beings within a sensibility that is always able to mediate their relationship with technology so that it is truly a *means* and not a substitute for the natural environment. To begin to get an idea of the plausibility and feasibility of such scenarios, we suggest examining in particular the reflections by Jaques Fresco (an American architect) in Peter Joseph's trilogy of documentaries: *Zeitgeist: the Movie, Zeitgeist: Addendum*, and *Zeitgeist: Moving Forward*, which can be viewed online for free. Equally congruous and illuminating are Silvano Agosti's reflections in his novel (*Letters form Kirghisia*) *Lettere dalla Kirghisia*, L'Immagine Edizioni, 2004 and in (*The invisible Genocide*) *Il Genocidio Invisibile*, L'Immagine Edizioni, 2008.

and environment) this is needed, 3) to assess what real impacts this would also have over the long range and in the long term, understanding how much one is really able to estimate and how much one is not, 4) to identify what is the best way to create what is decided (for the environment, for the users and producers of what is produced) by sharing the best know-how among all of us and 5) to get the products or services to those who need them through a way that creates the least possible impact. Just apply this to every human context and redemption will become a present as, little by little, generations of human beings will find themselves immersed in an increasingly collaborative and choral context in which the real needs of a species (human), embedded (and only possible) in its environment (the planet, *Gaia*, but in biological sense, not in the "New World Order" sense!), are met through biologically choices which make sense because they are driven by a great feeling of *care* and *connection*.

Before returning to our 'semi-synthetic life' *theme*, with what has just been said, we only wish to point out that, with the excuse of 'savings in the economy' on the one hand (that's a lie since the law of money is not a physical necessity but a pure convention from which we can – and should – set free all of us) and the 'new weapons in medicine' on the other (the problems of which can really only be solved by an epistemological matu-ration in biology and medical practice[48]), the obfuscation takes place of what is really being done and what implications it has on many other as-pects (such as the bio-eco-logical ones) that we must never, ever allow ourselves to forget. Attention is then shifted to the fact that what is being done would in the 'short term' give congruous and very probable ad-vantages and benefits (but always on the level of the categories that al-ready make up the social-technical-economic system, subscribing to it) and thus very few people pay attention to the fact that someone is design-ing and building semi-synthetic organisms (!) whose biospheric impact is possibly unmanageable.

The same "passing off" and substitution, beware, also takes place in the case of surveillance and control (from video surveillance, to water bot-tles to be thrown in the bin just few meters infront of the boarding gates at the airport and to be bought again a moment later, to taxation, to the traceability of credit card transactions, 'loyalty cards', motorway tolls and web chronologies) that are unquestionably prescribed (or even desired by

---

[48] See, for instance, sect. 1 (*Phenomenology of Perception*) *Fenomenologia della Percezione*, pp. 5-90 of (*We Are Our Body: Creatures of Perception Designed to Learn Unlimitedly*) *Noi siamo il nostro corpo: creature di percezione concepite per imparare illimitatamente*, by M. Sartorio, 2015, Amazon, sect. 1 'Phenomenology of Perception', pp. 5-90.

addicted users) under the high-sounding guise of 'security' or 'survey'. Any protest and intolerance to this 'necessary' intrusiveness would only be mystified and pointed at as a confirmation that one, then, has 'something to hide', and not taken as an attempt to preserve an albeit desirable sacred *intimacy*, in living our lives and in being in the world, which is increasingly compromised.

These substitutions, in which means and ends come to mix and mingle (or even swap among each other) as much as to diverge vertiginously on the level of necessity, testify to that *decay*, already precociously pointed out to us by philosophers such as Gunther Anders[49], so inherent to a modern-technical 'collective consciousness' in which specialisations become ever more acute and competences mutually isolated: one no longer grasps the difference between a task-*doing* and a conscious *acting*, science becomes technical and what one does is done because *technically* one can do it. That's all.

The "*doing*" does not imply any inter-relationship between the agent, the acted and the contours of the action (such as intellectual and epistemological premises, causes, effects, bifurcations, attractive processes or amplifying – positive – and dampening – negative – feedback mechanisms, etc.). For *doing*, a *technical competence* and the means to fulfil it are sufficient. *Acting*, on the other hand, is substantiated by self-awareness which, by necessity, is a *relationship between...*, it is *consideration of...*, it is *presence* in that of which 'wholly' one is participating. It is *critical thought*.

And so here is the big (but invisible) *problem* to which we liked to direct the focus above: are we in the presence of a *conscious acting* in science and co-science or are we in the presence of a *hybris* of the *technical doing* in which the relationships that characterise our *being living/alive* within, together with, a larger living *system* (*Life* on the planet) have been lost? The whole situation clearly indicates that the option is the latter.

But beware: in order to arrive at the *hybris* of *technical doing*, the consciences that implement it must first be sabotaged by a profound *removal* that pertains to a problem that is as much epistemological as existential: because that repression concerns a *quid* that dictates which the method of science is and which its criteria of truth are, and it is also the

---

[49] (*Gunther Anders: Critcs of technology and end of human*) *Günther Anders: critica della tecnica e fine dell'umano*, Francesco Miano, p. 201, in Hermeneutica, 2008.

one which codifies *how* we are in, *how* we *feel*, the world. It pertains to that feeling of reverence and sacredness we mentioned in the introduction.

It is necessary to consider well that such a removal afflicts scientific praxis as much in high-energy physics as in medicine, as much in agricultural techniques, biotechnology and genomics as in some postulated mechanisms in given approaches of sociology and psychology. In this sense, the problem of science which decay down to *techne* is the problem of the whole society. We are talking about a *mindset* that, ascribed to the (numerical and not only) categories of measurement, quantity, value and, ultimately, of logic, is stiffening the *breath* of all human gesture. And the totalitarianism of uniforming "rationality" offers its horrific spectacle without demeanour in the perverse marriage of the *total techne* with the *absolute capitalism*. We will elaborate on this theme in the penultimate chapter.

4. Perbenistic deliria
and epistemic narcissism

What is important to point out is that these critical reflections on an organised context, such as the one we experience every day at the level of societies, markets, economies, health, education, science and research, politics, etc., cannot be "respectably" dismissed in a ridiculed 'conspiracy theory' because there is no need for 'baddies' to be there to plan 'strategies' to put 'us poor, common citizens' in troubles. This is not what is being claimed here, although, in all honesty, one can no longer - in an uncritical and bigoted 'right-thinking' attitude - deny that actually 'some (human, financial, legal) subjects' have shown clear intentions and the power to direct many collective events, even on a planetary scale, at their own convenience. It is enough to think of the emergence of the large transnational corporations and the inevitable market 'agreements', or to consider how few are the owners of the companies that have in their hands the management of all the planetary broadcasting, or to see how often the majority stakeholders that have purchased large shares of social networks or global web platforms are the same ones that have large investments in some of the most well-known pharmaceutical giants, etc. The 'control room' can be there in a premeditated form or not. That is not the point. What is important is that in a globalised system in which the *human feeling* is replaced by the *calculating doing*, the generation of heteronomous dynamics with respect to the initial possible good intentions is almost a certainty.

The point is that critical thinking must be re-awakened to curb, to embank, such devastating *heterogenesis of goals* that undermine every expression of a communitarian living, corrupting it to a mere sum of interests and goals. The examples of this impoverishment even on more implicit levels are manifold: from the publication of experimental results to the obscuring of historical data, from the foraging of training content in education that constitutes a human terrain of poor critical capacity with regard to theoretical issues, to the deviation of entire economic trends as a result of banking agreements, trading conventions or ideological sympathies, without assessing what is good for all or only for a few. It is important, and realistic, to understand that many inhuman dynamics are also the result of *emergent processes* (and not necessarily of premeditated plans) that derive from the fact that *social poiesis* is distilled from what that can be dealt with in a repeatable, normed and reproducible manner, and that subjects can place on a collective (uniformed) ground.

Returning, therefore, to the problem of creating semi-synthetic organisms with a genetic code made up of six bases instead of four, we consider that science projects in which these are the goals are not inhabited by a *sacred sense of relationship* and that the operators who carry them out are no longer worthy of calling themselves 'scientists', self-degrading to mere technologists, or technocrats, depending on their hierarchical status.

Indeed, we must ask to ourselves: beyond the hoped-for practical advantages, what instances (and archetypes) drive some "insiders" to insert two extra nucleotides into a species' DNA? By which 'vision' of what biology is, of what Nature is, can one aim to the goal of constituting "semi-synthetic life" without considering the price of doing so? On what basis is it considered to be "a limitation" that there are four bases and not more? Why should it be better to "store more information" in a genome (provided it is true)? And what conceptions of *information* and *performance* (biological and not only) do underlie such a position on the issues at hand?

It must be restated: we are probably losing sight of something big that lies in the background. And that *something* concerns all, all the living systems. In fact, to those questions we can add many others on extremely sensitive issues: why should a mother (or a couple) who cannot have children (either for reasons inherent to the setting – e.g.,: homosexual couple – or for reasons of biological efficiency, in fertilisation processes) have them at all costs? Why do we not place ourselves in a perspective (humbler and more connected to our biological *selves*) that takes into account the *body's intelligence*? If one is not being able to procreate, it well means that there is a set of circumstances not suitable for that.

Why, in the grip of dissociative projections and heteronomous conditionings (speaking through an illusory "I want", instead of a deepened "I feel") that push us *to have to be* 'somehow equal to...', must we override the *bio-logical* (i.e., logical for life) *status* of what is there, capable of delivering us a sacred truth, if we remain open to listening?

How can we go so far as to say, with a hypocritical and pandering goodness, that 'lending the uterus' to someone who cannot procreate is an act of 'generous magnanimity', without grasping instead what a neurovegetative tragedy it is for a creature, a baby, generated and developed within the organism of a mother-female (also a victim), to be torn from it and assigned to other guys who (even if able to express wonderful good and care) are biologically extraneous to it (even though strictly genetically they are the "parents")? What an ugliness is it to implant a zygote given by the union made in a test-tube between the gametes of two partners and place it in a third uterus, as if it were an incubator...? Why does one not feel that with these methods of proceeding one degrades biological genesis, so magical and sensible, to something forced, dimidiate, even obscene...? If we were in danger of extinction, maybe (who knows?) we could even contemplate such extreme gestures, but to choose them because one has to fulfil a *arbitrary will* is simply a delirium and does not take into account that *intimacy*, a delicate and fundamental aspect of filiation, is being replaced by extraneousness.

It is an extraneousness that is not only macroscopic-physiological, but primarily *physical*, in the sense of the electrodynamic correlation states that are established between the molecular components and the *gauge fields* that mediate the interactions in that set of material and electromagnetic quanta that takes its onset as a 'mother-fetus' *unicum*. It is a unicum that begins and it is a unicum that remains, since those correlations are within the electrodynamic coherence that characterises the living phase of matter: as long as the vital state is in force, that long-range connection is performed, therefore the states (wave functions) of cells, molecules, genes and neurons are possible to undertake, and are subjected to, phase correlations (even non-locally) with the embedding in which they had their genesis and development (the biological, gestating, mother). But this is precisely the alchemical bond between a mother and her own child and it is of the same nature that allows the immanent self-correlation between all the representable 'parts' within an organism or ecosystem[50].

It is one thing to give care, cure and love to those who would have none (as in adoption) and who would be exposed to far more serious and by now irremediable inadequacies, it is quite another thing to determine/plan **a priori** an existence with the superficial illusion that there are no problems or differences with respect to what would happen according to processes developed by the biological default.

Another critical issue, therefore, is that of the parentage of homosexual couples. On this topic too, one has to ask: why do we "go outside" Nature and want to join at all costs two biologically incompatible actualities (i.e., filiation and the union of two subjects of the same sex)? Homosexuality is a natural fact and part of the balance of the species, among other things. And in such balances, in homosexual relationships it is

---

[50] See the first reference in note 19 and the first in note 28: Del Giudice, Brizhik, et al., 2005, 2009; P. Renati, 2021, *op. cit.*; P. Renati, 2022, *op. cit.*

consequential (logical) not to procreate: no children result from a homo-sexual relationship. And this is physiological, it is part of the 'pack-age/layout' in which consistency and coherence between biology and emotional-relational aspects is preserved. Also because a new-born child must be able to be constituted of both male and female archetypes and it is bio-logical that the setting in which life is generated represents them both through the corresponding, complementary corporeality-sexuality of the two parents.

Here we really must be very, very careful of both ethical, existential and psychic, and biological simplicisms. Such simplicisms undermine, in-deed, even the assessments about the "rightness" (and lawfulness) or not of the adoption of children by single individuals (including heterosexuals) or homosexual couples (or those with "other orientations", or with gender fluidity, etc.). It is not, in fact, a question of the capacity or incapacity of 'giving love', which obviously does not pertain to sexual orientation at all, but rather a question of the *quality* of love, in the sense of the *typicality*, of the *modus*, of its 'structural coupling' to a given sex-implied *corpore-ality*: the way of experiencing and giving love of a woman-mother is not the same as that of a man-father, for instance. Which, fortunately, is func-tional and necessary.

The central point, then, always lies in grasping the nuances and dif-ferentiations: that is, one must recognise that biology, life, about species that reproduce sexually, has evolved a composition of complementary and co-functional *qualitates* that act to make up not only the somatic expres-sion of the subject, but also the emotional-archetypal one.

Therefore, it is simplistic what, for example, the esteemed and eman-cipated theologian Vito Mancuso maintains[51], regarding his position in fa-vour of the adoption of children by homosexual couples, by saying that «love is all that matters», thinking of such «love», in reference to a hypos-tatised god, as something decoupled from Nature. This is the great prob-lem of the ontological dualism that still plagues contemporary culture (both secular/lay and religious). In fact, modern man (the one who, pre-cisely in this removal of the "transcendent" from the phenomenal, has pro-duced an "alienated Second Nature" [Adorno] of calculation and reifica-tion) has a great "lesson" to learn from the observation of biological sys-tems, i.e.,: the teaching that allows us to overcome once and for all the dualism between *being* and *nature* (still unfortunately upheld by theology,

---

[51] Radiophonic interview on the programme (*The whole city talks about it*) *Tutta la città ne parla*, RAI-Radio3, episode of 10 March 2017.

despite the fact that B. Spinoza has already shown us their identity[52] in an indefatigable and unbreakable manner), by showing how that *love*, like any other dynamics, is nothing detachable from its own *structural coupling* (to use Maturana and Varela's terms[53]), i.e., the *soma*, and is nothing *other* than it: in biology, the isomorphism between *medium* and *message*[54] is in full force.

This has already been well demonstrated to us by the neuroscientist A. Damasio[55], in having confirmed that every feeling, and therefore every emotional action, is nothing more than a precise visceral configuration (pertaining to the neurovegetative system) in which the *emotion*, the *feeling*, takes root precisely as a *somatic marker*-state. Otherwise it would just be a thought. But we know very well that the mere thought, the concept, and therefore also an aseptic "verbal statement", does not involve an emotion that is accompanied by a whole vegetative (oscillatory) condition, which is also perceivable by the other always and again neuro-vegetatively (and not cognitively!).

With what has been said, then, we want to highlight how acting out the feeling of love (actually, some form of it) cannot be decoupled from the physical corporeality and *biosis* of which it is an expression. In saying that "love is enough" one commits a terrible simplification that does not consider how a born human being needs to breathe those precise *qualities of love* that give each other the counterpoint in what biology has evolved in those characteristics of the *male-female* couple. The ancient Greeks have already taught us that love is not all the same and that even its, if you like, highest form, *agape*, is not enough... in the sense that it is not what is always needed (or not only). To give a simple example: if we say we love our partner, we cannot think of acting out a sentiment such as that which can move the actions of a Franciscan friar in a mission in Africa, or even the one which we give to a brother. A precise form of love, in the case, *eros*, in that type of relationship is necessary, i.e., contextual, for the feeling of 'love' to be authentically expressed and truly 'felt' as appropriate and satisfying.

In the same way, a subject (especially an infant) needs a *concert* of these *qualities* in which *eros* (passion and sexuality), *philia* (dedication

---

[52] "*Deus sive Natura*", see: Definition 3, in *Ethics*, Barruch Spinoza.

[53] Maturana, H.R. & Varela, F.J., Original title (*De maquinas y seres vivo*, 1972), *Of Machines and Living Creatures, Autopoiesis, Organization of Living* (title adapted), Porto Alegre: ArtesMédicas (1997).

[54] *Understanding Media: Extensions of Man*, McLuhan, M. MacGraw Hill: New York, NY, USA, 1964.

[55] *Descartes' Error: Emotion, Reason, and the Human* Brain, A. Damasio, ed. orig.: Putnam, 1994

and involvement), *storge* (brotherly or family love), and *agape* (love without identification, panic and universal) unfold and flow with the hues given by the archetypes of the *male* and *female* harmoniously coupled to the corresponding corporeality. A girl also learns to express her femininity through the eros of her parents, of the introjected models, and makes this analogically. She cannot attend courses. Are we sure that nothing would change if, as reference models, she only had two parents of the same sex?

The matter here is not about heeding studies whose statistics lead to the assertion that, on average, children brought up by homosexual couples show no difference from those brought up in heterosexual families[56], or those studies whose conclusions are the opposite[57]. Rather, it is a matter of pointing out that in those studies there is no place but for 'descriptive' observables, the so-called *well-being measures*: school performance, cognitive development, social development, psychological health, sexual precocity, substances abuse, career probability, etc.

And who looks at the more intimate sensations (from 'skin, belly...')? Who considers what the neurovegetative memory of a contact with the physicality of the biological mother-female, biological father-male provides, and how these complex perceptions are integrated in the new-born and how they will translate into the adult regarding the recreation of states of comfort, deep identification, discernment of pleasant/unpleasant stimuli and the development of 'bodily competence' to be able to choose healthily, consonantly, where, how and with whom one *feels* comfortable ('in the belly') and *safe*...? No one ever deals with the archetypes and their structural coupling to the body, and we hide everything in statistics, forgetting that "on average" can mean very little, since there are so many factors that intervene on those variables observed by the "measures" of the aforementioned studies.

Many, in fact, like to think that, as some of the aforementioned works show, the consequences on the development and self-consciousness (also

---

[56] *Child Well-Being in Same-Sex Parent Families: Review of Research Prepared for American Sociological Association Amicus Brief,* Wendy D. Manning, Marshal Neal Fettro, Esther Lamidi, Manning, W. D., Fettro, M. N., & Lamidi, E. (2014). Population Research and Policy Review, 33(4), 485–502. http://doi.org/10.1007/s11113-014-9329-6, https://www.ncbi.nlm.nih.gov/pmc/articles/PMC4091994/pdf/nihms-594603.pdf

[57] *Same-sex parenting and children's outcomes: A closer examination of the American Psychological Association's brief on lesbian and gay parenting.* Marks L., Social Science Research, 41(4), 735–751 (2012). *How different are the adult children of parents who have same-sex relationships? Findings from the new family structures study.* Regnerus, M. Social Science Research, 41(4), 752–770. (2012). *Parental same-sex relationships, family instability, and subsequent life outcomes for adult children: Answering critics of the new family structures study with additional analyses.* Regnerus, M., Social Science Research, 41(6), 1367–1377. (2012).

bodily) of children brought up in homosexual couples are ultimately un-differentiable from the biological default, but it has to be said that, actually, in the data that would confirm this, one can recognise a homogenising factor due to the fact that in both cases the children can still experience other models within the social and family context, in addition to those of their parents. The point to be made here is to pay attention to what one observes in order to decide whether one sees differences or not. Researchers are strongly advised not to limit their observations to behavioural or merely neurological aspects, neglecting then what constitutes the entirety of the subject's unconscious perceptive (and archetypal) structure[58] (which, in fact, preponderantly orients his existence and self-consciousness). As one can see, the flaw is always the same: how little relational and systemic one observes reality in a self-styled "scientific" approach, which is in fact simplistic insofar as it is partitioned.

To be fair, even the 'conventional' family model of (not geographically speaking) Western society with parents and children, confined in their own private context, is not totally bio-logical and has been structured under the deforming pressure of the organisation of space, work and of roles superimposed on the biological essence. The *Homo Sapiens* animal is notably social, and, like any other organism, it can experience, and express itself within, a *range* of reality that is comparable with its emotional (perceptive) scope and its own spatial-temporal scale. The non-biological extremes in which we live in the contemporary layout have both materialised: on the one hand, immense masses of individuals who are relationally disjointed, but informationally in (intrusive and pervasive) communication; on the other, the annihilating isolation within the loculi of their own double-enclosed flats within which 'families' are housed.

Yet, human beings would develop far better in a human-friendly setting, such as in *villages* (of whatever culture), where 'everyone is everyone's child' and everyone has regard and care for everyone else, and where children can experience so many variations of mum, dad, male and female, and 'orient themselves' according to their own feelings on how to express those archetypes within themselves. In the blocks of flats in the cities, even the dear old *courtyards*, in which families shared part of their existence, are now precluded, and also the squares and streets in which the activities (and idleness) of adults and children created a concert of multipurpose relations functional to the *formation* (*Buildung*) of the life of the 'human puppies' (with their games in the street among artisans in the

---

[58] *The Relations between the Ego and the Unconscious*, Carl G. Jung, 1928.

workshop, the women talking, the elderly at chess play) are now scarce. In the cities today, it is not uncommon to meet young people sitting on benches or pavements around a smartphone, suggesting disparate contents that are as seductive as inhuman. These include stereotypical representations of success, intelligence, love and sexuality. And equally stereotyped is also the transgression of such 'models', however codified in eccentricities that become clichés.

Another important aspect within the human existential horizon, which sees an obvious degradation in the transition from an anthropic model within human reach (tribal or village) to a metropolitan model, is that of the biological expression of sexuality: illuminating in this regard are Malinowski's studies on the populations of the Trobriand Islands[59], from which it can be deduced how a balanced and bio-logical experience of one's sexuality, from the earliest years of life, within some tribal contexts (not over-structured by prohibitions by moral or heteronomous norms in general) leads the subject to a healthy existence not conditioned by removals or traumas and conflicts between the intimate *felt* and the ethical *allowed*, externally expressible.

We may also recall what Silvano Agosti has already well pointed out[60] about the fact that tenderness (*philia*), sexuality (*eros*) and love (*agape*) must always be able to go hand in hand, especially in the love of a couple, otherwise: tenderness without sexuality and without love produces *hypocrisy*, eros/sexuality separated from tenderness and love produces *pornography*, and love isolated from tenderness and sexuality produces *mysticism*. And in fact, our western society is precisely characterised by great *hypocritical* drifts (about the presumed irreplaceability of models of development and organisation), *pornographic* drifts (about the replacement of subjects with images of subjects and emotions with their stereotypes increasingly shared, made public) and *mystical* drifts (as in the theological case in which everything is sublimated into a quintessential "love", which can be freed "from the flesh of the body", subscribing to an ontological dualism between the world and its own metaphysical causes).

At this point, regarding the "ethical" aspects of the problem of children within homosexual couples *et similia*, it would be good to distinguish very clearly three fronts.

---

[59] (*Sex and Sexual Repression among the Savages*) *Sesso e Repressione Sessuale tra i Selvaggi*, Bronislaw Malinowski, Bollati Boringhieri, 5th edition, 2013.
[60] (*The invisible Genocide*) *Il Genocidio Invisibile*, Silvano Agosti, L'immagine Edizioni, 2008.

(i) One concerns the fact that a child, thanks to adoption, can be removed from traumatic and damaging realities; this is unquestionable and welcome whenever there is a need, but it is not the point at issue.

(ii) It is another matter to deliberate that the conditions in which the adoptee would find him/her-self are equal and equally suitable with both heterosexual parents (as per biology) and homosexual parents (who, as per biology, are not parents, though they may be exquisite, generous, capable of great feelings and sacrifices and/or excellent pedagogues); and that is what is being discussed here.

(iii) The third point concerns the devious mystification of those who believe – puritanically – that "we are all the same" (incorrect, since, if anything, we are all different and equally dignified!), or that "we are all human" (which is not in doubt at all!) and who mean that acknowledging such differences, which are self-evident and exist in themselves, in an act of "discrimination".

'Double standards' are being used in this regard. It is truly absurd that, for example, one can define as 'sick' a subject in whom there is an adenocarcinoma on a certain endodermic tissue, without realising that this is a response with a precise biological meaning[61], but that one does not instead recognise that behavioural (or even hormonal) variations associated with bisexuality, homosexuality, trans-gender, hermaphroditism are themselves also psycho-neuro-biological results acted out as adaptive responses to certain configurations of the environment (especially systemic-relational and familiar, especially of the subject's early stages of development). No one of such situation should be defined as a "disease", actually. And this not by "linguistic bigotry", but because we understand how neurobiology works[62]. Someone will certainly object that an adenocarcinoma can lead to death, whereas homosexuality, hermaphroditism or feeling like a woman in a man's body cannot. And that it is therefore legitimate to call the condition of the former a 'disease' and not that of the latter. But this assertion is always the result of a lack of knowledge (and voids in

---

[61] Cannon W., op. cit.; Hamer R.G., op. cit.; Maté, G. (2003). *When the Body Says No: Understanding the Stress-Disease Connection.* Hoboken New Jersey: John Wiley & Sons, Inc.

[62] Porges S.W., op. cit.; *I meccanismi Neurobiologici della DHS*, D. Toneguzzi, Psiche Cervello Organo 1/2006, www.biophysics-research.com; *Il conflitto Biologico*, D. Toneguzzi, in Psiche-Cervello-Organo n° 1/2007, www.biophysics-research.com; Hamer R.G., op. cit.; *Manuale di Applicazione delle Cinque Leggi Biologiche (scoperte dal Dott. Hamer) Vol. 1, Svegliarsi dall'ipnosi della "malattia"*, Marco Pfister, Secondo Natura Editore, 2013 ISBN 978-88-95713-17-5; *Comprendre sa maldie*, Michel Henrard, Ed. Amyris, Bruxxelles 2015, Ed. It. *Comprendi la tua malattia – con le scoperte del Dottor Hamer*, Macro Edizioni, Cesena, 2015; *Psico-Bio-Genealogia. Le origini della malattia*, A. Bertoli, Macro Edizioni, Cesena, 2010, ISBN 88-6229-156-6.

verifying what one believes, or what it seems to us) of the biological functioning in which the cause of death is not given by the activation of a response that is in itself physiological (sensible and by default aimed at survival) as an adenocarcinoma can be, but, if anything, by the recurrence/perpetuation of the conditions that trigger it and/or the undertaking of actions (sometimes also called 'therapeutic') that are counterproductive because they are conducted on an epistemological basis that has not accepted the body's intrinsic intelligence and the living being as a perceptive process[63].

So let's be honest, if homosexuals or trans-genders are rightly not called 'sick', or it is said that those who have certain conditions that 'differentiate' them from the *biological default* are 'differently abled', then neither schizophrenic, autistics, bipolar or trisomy-bearing people should be called 'sick'. After all, they too are specificities that cannot be considered solely negative in an arbitrary manner (certainly, provided one has abandoned a certain hypnotic view of 'illness'). Then, however, may such specificities be factually (and not only in words) acknowledged with regard to the fact that, to each of them, one must adapt with common sense, case by case. Who can deny, for instance, that autism also implies special sensitivities and faculties, precluded to the non-autistic subjects? Can we call the former 'differently abled'? Let it be. But we know that this 'different ability' implies being better at doing some things and being less able to do others. Can we have a concert for twenty thousand spectators organised by a trisomal boy? Quite not. Can we entrust a class of ten kindergarten children to an autistic person? Quite not. Fine, but if this neutral definition of factuality, in the recognition of differences, is condemned as discrimination, then we are in bad faith, bigotry and inconsistency[64].

---

[63] Toneguzzi D., op. cit.; (*We Are Our Body: creatures of perception conceived to learn unlimitedly*) *Noi siamo il nostro corpo: creature di percezione concepite per imparare illimitatamente*, M. Sartorio, 2015; P. Renati, 2021, *op. cit.*; P. Renati, 2022, *op. cit.*.

[64] The distortion, then, often reaches acute peaks when even to the *differently abled* are given *jobs*, *tasks* as a sign of equality, as an offering they have to feel honoured of in some way. It certainly makes sense that in the process of *integration* within the *social* context, we aim at the *inclusion* of (also) differently abled people within everything that involves them in human relations and makes them feel endowed with meaning. But once it has been accorded that, by right, every means for living with dignity must be guaranteed by default (without asking for anything in return) at least to those who have additional difficulties due to disability, we should not make the mistake of tracing the meaning of existence, and thus also *human dignity*, to the category of *work/job*, i.e., to being productive or socially useful. Is it the case that one is democratic even about the great deceptions? Work as a "right"- the great lie of the culture of capital - also becomes an aspiration, because if you work you are "fulfilled", they say. Where else has the subscription to an alienating system that makes one believe that one can only be in the world if in subordination to an economic-productive identity (and if one therefore 'earns money' - by working), reached such apogees? The organisation

Let's say "stop!" then, to a *'politically correct'* respectability that prevents us from recognising things for what they are, dryly, without ideological *bias*: in the same way as saying that this is an avocado and this is an apple, that this gentleman has a limp and that one does not, that this girl is on a wheelchair and that boy runs and plays football, that this is a poplar and that is a carnation flower. To each structure its own expression and function. And therefore its own possibilities. Simply. Serenely, in acceptance.

Otherwise, even any act in which differences manifest themselves even in common circumstances would become discriminatory: it would have to be an immoral outrage when a gymnast's performance is seen by paraplegic spectators, or to enjoy a mime scene in the presence of blind people. Expressing our own faculties, skills, typicality, would be an 'insult' to the 'limitations' of others 'who cannot'. Yet, in 2020, at Harvard University[65], they come up with closing courses about the Greek Classics, because – it is said – there were only white-skin people and this 'discriminates' black people (!).

---

of work today has anything but emancipatory connotations given the impossibility of doing without it if one wants to be entitled to a livelihood. A fact, this, that sabotages *ab initio* the possibility of expression by vocation and in accordance with ways, times and rhythms that are chosen by the *feeling* of the subject. It would be an anthropological conquest if at least those who already have difficulties due to disabilities did not have to enter the theatre of "work as a producer of meaning": human relations and expressing oneself with others (to be promoted and enhanced in increasingly affective and existential aspects) is one thing, flattening inclusion within systemised tasks is quite another. This only underwrites one of the most serious alienations of contemporary society. Moreover, and more importantly, *sense* and *dignity* are not to be traced back to any productive role, whereas they are to be traced back to the *essence* of the living (and human) being itself. About this, compare the in-depth analysis by Barbara Malvestiti, in (*Human Dignity after the Nice Chart. A conceptual analysis*) *La dignità umana dopo la Carta di Nizza. Un'analisi concettuale* Orthotes Edizioni Publisher, Napoli-Salerno, 2015, in which social dignity is consistently revealed only «as a particular case of *human dignity*» (pp.104-105) and with regard to the dignity of the disabled person it is written (pp.110-112): «Respect for dignity of disabled person coincides with the protection of his/her person, with the prohibition to reduce it to the datum of his disability, offering conditions for its flourishing. There are different levels of flourishing. In some cases, the need for flourishing can be exercised at a minimal level, which does not require autonomy, but rather an affective relationship with the world. In other cases, the need for flourishing implies autonomy and involves one's own life and value choices. It is at this second level that the most widespread understanding of the concept of human dignity as the dignity of the individual imposes itself: human dignity as the *protection of autonomy*, i.e., the possibility of determining one's life on the basis of personal preferences and convictions». In our opinion, therefore, this fulfilment of the individual's *dignity* and *autonomy* - as the possibility of self-determination on the basis of preferences - is all the more guaranteed and authentic (in privilege of a fundamentally *affective* experience of the social) the more the subject is inserted in a human context whose sensibility has already allowed, on the one hand, emancipation from "total organisations" (with Heidegger) and, on the other, transcending the sense of the subject understood as subordinate to his being useful to some productive purpose.

[65] https://www.ansa.it/sito/notizie/cultura/libri/approfondimenti/2021/04/20/luniversita-nera-cancella-i-classici-polemica_9d0dac96-e5e2-40ab-b46a-7b655b25f487.html

But, in a nutshell, why can there not be human beings who in their joy and peace of mind do not have children, as even aware that they do not correspond to the elements of a biologically (and perhaps momentarily) suitable setting for that experience[66]?

What we are talking about, in terms of the weakening of the capacity to recognise the real biological *requirements* for procreation and nurturing, in which behavioural and sexual models keep cogency and necessity, concerns precisely this kind of distorting substitution between what is biologically consonant insofar as it has evolved in coherence with the functions of *biosis*, and what is arbitrarily, ideologically, possible as a technical feasibility. The same distortion that occurs in implying that the human dignity of differently abled subjects finds due realisation and recognition the more they are included (by conforming them to the same interactive models) in social activities, including making themselves useful. We find that this vision, if at first sight it may seem emancipating and egalitarian, hides a risk typical of the technocratic perspective in which individuals become more and more dignified in their existence the more they 'serve something'. Let us explain ourselves further.

While, for example, it is true that an autistic subject may have incredibly better calculating and visual skills than a 'normal' subject, and that they would even be expendable in some practice, it is certainly not desirable to be autistic! The fact that a 'special' circumstance exists (such as a child having only one mother, or being raised by two men), does not mean that it should be substituted, normalising it, for the *natural default* that makes sense and is logical for life under optimal conditions.

This is precisely the same problem that afflicts synthetic biology, conceived as the manipulation of biological systems to obtain purposes: what purposes? Purposes that are not already envisaged within the default of the biological *logos*. Otherwise, they would have been evolved (or in the process of being evolved) along the phylogeny. So, if, paradoxically, we were to develop a social system within which the faculty of being able to count in a blink of an eye how many matches have been thrown on a table is expendable or empowering (think of the scene in the film *Rain Man*, starring Dustin Hoffman and Tom Cruise), would it make sense to

---

[66] Above all, because an important difference is recognised, as Simone Weil already pointed out, between mere *desires* and the *needs of the soul*, emphasising the difference between an *accidental* need for the former and an *essential* need for the latter, inasmuch as only the latter are vital, that is, «nourishment for the life of the soul»; cf. S. Weil, *L'enracinement. Prelude à une Declaration des Devoirs envers l'Être Humain*, pp. 12-14 in the Italian translation (*The first root. Prelude to a Declaration of Duties towards the Human Creature*), *La prima radice. Preludio a una dichiarazione dei doveri verso la creatura umana* Milan, Edizioni di Comunità, 1954.

willingly promote autism through genetic techniques? The answer, before being ethical, is, biologically, a clear 'no!', but if one were to stop at the objectifying gaze admitted by technical expendability alone, such an answer could be a cynical 'yes', since it could be aimed at some purpose, productive (and therefore heteronomous to the subject itself). And being autistic would become an advantageous condition for that 'purpose'. The problem, as one can well guess, is about 'the rest' of the person's expressive and perceptive horizon: how is it changed and what consequences does this change have? Obviously if autism were to become the default human being, we would have serious problems configuring a functional and coherent social semantics. But we know this, and yet right here lies the problem of wanting to pursue (arbitrary) 'purposes' outside the biological way.

Well, in this *deliberate will* of Heideggerian denunciation resides the same perversion that considers organisms whose genomes are composed of six nitrogenous bases, instead of four, as possibly 'useful'; or that considers the creation of GMOs and/or hybrids of corn or wheat more suited to certain conditions or purposes (e.g.,, resisting pesticides or drought, or producing more) as 'good and useful', etc… All this always without taking into account that the need to fulfil those 'purposes' arises within a given setting (in the case: a certain way of farming or 'making medicine') that is far from being unchangeable and far from being the best possible; hence it must be questioned.

In the same way, a homosexual couple can certainly give love and raise a child, just as a child can be conceived in a test-tube or in another womb, for all practical purposes, but – let us ask ourselves – is that a situation equal to the *biological default* or even to be decided willingly? We should say not, especially by considering the relational and archetypal aspects we discussed above.

While we are perfectly aware that identitarian definitions provide back to us almost nothing about reality, let us point out that, although we know that 'normal' means nothing in particular, in nature it means something in general, so much so that, if one is able to recognise an antelope, a human being, a cuttlefish or a maple tree, it is thanks to their specific morphological and dynamic characteristics "as normal", because they have a biological *default*. This fact is well verifiable. When, for instance, in the human being one has to deal with a postural or orthodontic 'defect' (always a consequence of an adaptation that sees that reconfiguration as a strategy which makes sense) and one finds, as a cascade, changes in the efficiency of certain movements (in terms of amplitude and force), one recognises this because there is a biological 'normality' of reference (and one tries to relate the structure-function of the system to it, whenever possible).

In the current drift, in which both science and religious confessions and adhesions of thought, break humanity down into splinters oriented by different beliefs, we witness a *hybris* that, however, is translated on the level of praxis in a homogeneous and cohesive manner since it all feeds a pervasive attitude that (more or less deliberately) almost aims at overturning what the connectional network of the *bios* has laid down with intelligence, progressive refinement and *sense*. This *hybris* is the same, albeit originating from different 'factions' and 'representations', on an ideological level even antipodal, because it pertains to the common root of the problem: an identity-based, logical, causal, local and finally *dual* thought. The same bug that has degraded science to technology and spirituality to religion.

And it is really surprising that, jointly to the fact that technique is more and more allowing increasingly extreme bio-medical practices and various manipulations, a lot of anthropological and cultural motivations bloom out (expressed through the forms of struggle for equality, demand for equal rights, emancipation, such as adoptions by homosexual couples, or procreation by individuals) that go precisely in that direction, endorsing this technocratic hegemony of calculation and manipulation (on the epistemological side) and of eidetic representation and uniforming/standardizing belonging, affiliation, membership (on the psychological-social side, of "wanting to be like...someone/something").

Everything, from science to religions, from cultural customs to the corresponding jurisprudence, is moving towards an ever more irremediable distancing of man from his biology and from his healthy relationships capable of giving meaning to feelings and life only if they are lived, respected and felt.

Such a madness will remain irremediable until, at an anthropological level, the way of thinking of human beings is integrated, and in fact revolutionised, by an *analogical sensibility*, the only one capable of contemplating the dialectical co-participation of the fundamental oneness underlying the phenomenology of reality and the infinite differences that give to it *form* and *action* (in-form-a[c]tion).

Differences, taking up the thread, have equal dignity, but precisely as such, i.e.,: differentiated, distinct, diverse. That is to say, a cat has the same dignity as a dog *because* it is, and *yet* being, *different* from the dog, and not because the former is to be traced back to attributes *proper to* the dog.

We must learn to recognise that the cornerstone of an authentic democracy should be the equal *dignity of differences* and not only (and not so much) the *equality of rights*, especially if within the definition of "right" end up aspects that have nothing to do with it.

In fact, the whole distortion underlying the issues on procreation and adoption discussed so far (from uterus renting to artificial insemination, to children raised by homosexual couples), starts from the **great misunderstanding** that "having children" is assumed to be a *right*. But not at all! This false identity between procreation-caring and rights reveals how the perspective from which such experiences are viewed is that of (already conditioned) adults who want to 'realise' themselves (whether heterosexual or homosexual, or whatever) and not that of those who will be born. To say that "procreation is a right" subscribes to the fact that there is someone (already alive) with *needs* and who therefore "has the right" to satisfy them (through who will be born).

But when parenthood is meant to be a *need*, already the first defeat, the first distortion of that sacred experience occurs. And herein lies the great distortion that afflicts the existences of the alienated human being (not only today): too often children are generated to fulfil purposes or to fill existential gaps and shortcomings that ended up in producing *needs*: needs for love *to be received*, needs to feel socially adequate, or, as in the countryside in previous decades and centuries, needs for human resources, labour power, i.e., economic ones. Then, the fact that all of this is accompanied by immense emotional involvement and great experiences in which wonderful feelings come to unfold, is not denied by anyone. But let it be clear that, as long as one thinks that "one has children" and that it is therefore "a right to have them", one is greatly misguided, prey to attachment and victimhood. Children must not be *sons/daughters of needs*.

Speaking, then, of love, as for example Mancuso did (however intellectually and ethically estimable), not only is it necessary to specify the archetypal colourings and *qualities* and the *structural couplings* to the corresponding corporeity (mother-female, father-male), but also the direction: when children become the *instrument* (albeit unconsciously) of self-proclaimed "parents" in order to fulfil their desire to feel "like the others" (adequate), then those children are already born with a (love) debt, and the love spoken of is not the one *to be given* (unconditionally), but the one *required* by them. Right like that 'love' that is also expected and drawn from the social (and parental) context in the form of approval and recognition, congratulations, a sense of belonging/membership... these are all dynamics that pertain to an unresolved narcissism and to a profound un-freedom.

Please, let's comply with biology, where only a few (neuro-vegetatively) *chosen* components reproduce, and let's recognise that procreation and nurturing are experiences that are *encountered* along our lives and not "decided". We do not have children; we *receive* them if there is a structural coupling and a biological semantics consistent with such an experience. It is biology to set this. And this applies to women, men, hermaphrodites, trans-genders, lions, trees.

Homosexuality must obviously be accepted as equally dignitary to heterosexuality and any other consensual and peer-to-peer expression of one's sexual orientation. And what must be accepted and put on an equal level are precisely the *differences* and the implications that follow. So let us be coherent and recognise ourselves fully for what we are, accepting all that such a condition inherently (and not artificially) entails.

An appeal is made to those who recognise themselves in various sexual expressions, all of them dignified and equal: accept and welcome with dignity, honour and coherence one's own characteristic nature, one's own specificity, with all its interesting and non-trivial sides, and do not fall prey to narcissistic representations in which one feels "less than…" because one is confronted with "other". In such gesture one is not welcoming oneself in his/her own dignified difference, one is not loving oneself in what one is. Rather, one is evaluating oneself and looking at oneself through a collective filter, through an impersonal objectivising gaze that, for the most part, represents the average social thinking, hence that of the majority of people who have a given (average) type of life, with some experiences, shared by most, and so on.

But one has to be able to say a healthy "who cares!", definitely freeing oneself from conformist conditioning. If one feels inferior because in one's biological condition one cannot have children, one takes the same blunder as a leopard that, not being structured to fly, would like to resemble a condor.

It is the same great deception that many women have told themselves regarding that they would have been of "equal value" to men if they could spend themselves in the same activities that the latter do. And so today we have that even the (proud) girls go to the front to shoot (and to be shot). If there is something that is immensely jarring, given the innate embodiment of tenderness, acceptance, and participation typical of the feminine, it is precisely the "woman-war" pairing, yet… it is "emancipation"! This is truly grotesque… as if a man felt denigrated by not being able to wear a pink tutu for the ballet at La Scala. But differences are so beautiful, (especially between female and male), let's not flatten them!

To play a deadly role in these delusions is the blackmail of 'work-economic autonomy'... and so to care of the children there are grandmothers or nannies because "I am an autonomous woman, not <u>only</u> a mother". Those who think like that risk to be already distant from the whole *sense* and *wonder* of the experience implied by being a woman-mother, with all the uniqueness and totality that such a condition represents in relation to life and one's generative role[67]. *Proximal dropouts* abound and the consequences within twenty years of this aporetic paroxysm between parental and professional roles can be seen very well: youthful nihilism, in fact, starts first and foremost from these emotional holes and from the consequent attachment patterns (insecure, anxious, disorganised, etc.) that result. All this does nothing but consolidate the atrophy of the teleological and empathic faculties: precisely the terrain suited to root an epistemological, scientific-technical, and social paradigm such as the one that underlies the *uncon*-sciences we are narrating.

Unfortunately, when subjectivities are already dimmed by expectations, beliefs and collectivised representations in which the social anonymity decrees who is successful and who is not, who is 'cool' and who is not, who is admirable and who is not, an (almost) irremediable estrangement is sanctioned from the possibility of re-perceiving oneself within that great biological and archetypal *connectome* so delicately subtended to one's *true* existential realisation. And mind you, this does not at all mean that every woman (or every man) must be a mother (or father). On the contrary: what is being said is that the parental experience should be an authentic *calling* acted out by self-consciousness and strong selection and, once such, one should know how to put it before all the junk and dissipation in overlapping roles that subtract energy and time from *something* that requires infinite *care*. Since when we bring a *new life* into the world, we owe it all of ourselves in terms of listening and emotional nurturing.

The problem of semi-synthetic life lies on the same trajectory as the denaturations just discussed, with which the issue of artificial insemination (an, today, also that of ectogenesis) and GMOs are equally associated. How can anyone think that an embryo created by the joining of gametes in a test tube or under glass is a life that lacks nothing? Why don't we observe the phenomenology of biosis and realise that the act of conception

---

[67] See: (*When we were females*) *Quando eravamo femmine*, Costanza Mariano, Sonzogno Editore Publisher, 2016.

is the first true 'act of birth' in which all the energetic, emotional, existential and relational premises of the mating partners go (in addition to the recognisable biochemical composition) to shape that 'excess' energy necessary for the emergence of life through the reflex of orgasm? It is a matter of taking into account oscillatory arrangements and long-range phase correlations that determine the neurovegetative condition at a level below biochemical and receptor mediation. The rhythmicity of the system is, simply put, the state of its oscillations, and thus its *modus*, its semantics and its emotional-perceptual set-up, at that moment. All biochemistry is produced as the output and mediation of such electrodynamic premises[68]. A test-tube embryo, whether we like it or not, lacks much of this 'oscillatory coupling and *embedding*'. This is not meant to be a statement for discriminatory purposes; on the contrary, it serves to draw attention to *evident* and unavoidable aspects in the physics behind *biosis*, albeit not directly measurable, and serves to refine, mitigate, our technological gestures.

Moreover, when assisted fertilisation is motivated by 'demiurgic' (or eugenic-like) intentions with regard to the genetics of the baby – i.e., when such a practice is joined to genetic *editing* techniques aimed at "moulding" given characteristics (aesthetic or functional) – the problem is worsened by a further conflict between *self-determination* and *autonomy*[69], which fortunately someone has already analysed theoretically, proposing a solution in favour of the latter. Quoting again from B. Malvestiti[70]: «According to Habermas, in the case of a conflict between self-determination and autonomy, a liberal society subordinates the former to the latter. Thus, for example, in the case of a conflict between the parent's psychological need to determine the child's genetic make-up to satisfy aesthetic needs on the one hand and the future individual's (the child's) need to feel responsible for his or her own life choices and values on the other, a liberal society puts the latter first»[71]. A "primacy of self-determination" risks a subscription to a "deliberate (self-imposing) will" (about which Heidegger alerts

---

[68] P. Renati, 2020; P. Renati, 2021; P. Renati, 2022.

[69] This distinction has already been illustrated by S. Weil, as reported in footnote 63 when distinguishing between accidental desires and essential needs of the soul, Weil S. op. cit. In B. Malvestiti, op. cit., ch. 2, p113, it is explained as follows: "According to this distinction, while autonomy implies the level of personal investment that the individual typically places in his life choices and values, self-determination implies the minimal level of psychological needs".

[70] B. Malvestiti, op. cit., chap. 2., p. 113.

[71] See also Jürgen Habermas, Die Zukunft der menschlichen Natur. Auf dem Weg zu einer liberalen Eugenik?, Suhrkamp, Fankfurt am Main, 2001; Italian translation by Leonardo Ceppa, (*The future of human nature. The risks of a liberal genetics*) *Il futuro della natura umana. I rischi di una genetica liberale*, Einaudi publisher, Turin, 2002.

us in the passage quoted at the beginning) that, basically, is the same un-consciously despotic attitude of *technological man* who acts in substantial solipsism with respect to the *natural womb* of which he himself is in truth a generation, not caring about 'greater relations' and having as his only limit that of technical feasibility.

Technical feasibility is a hinge that becomes in truth also the *primum movens* for that "identitarian making" since it almost finds the sense of its own fulfilment in the self-demonstration of its own power. The fact that technically one *can* do something becomes the motive for implementing that something and at the same time constitutes (temporarily) the maximum limit of what one *can* do (a limit that one looks forward to transcending, regardless of a vision that has the systemic preservation of the organism-Nature as its firm point).

5. What is *life*?

The original breadth of meanings, senses and qualities proper to the vital process in Nature undergoes a tragic collapse at the hands of the epistemological reductionism that characterises the bio-technological practices discussed so far, which reduces 'life' to a mere 'molecular condition', a mere 'biochemical recipe'. In fact, to consider a frozen or 'mounted in a test-tube' zygote as *endowed* and *connoted* in the same way as an embryo generated by a mating 'according to biology' is a fictitious identity based on the slippery belief that we are only dealing with a "soup biochemically adequate to function". A stand, this one, that is always enslaved to the illusion that what is "seen" or "measured" is able to count for the whole system in exam. Modern molecular biology has given us a merely structural view of *living matter*[72], ignoring what 20th century physics has incontrovertibly brought to light: the dynamic aspects. The functioning of biological matter within the mainstream of science is in fact supposed to be integrally representable through electrostatic interactions, key-lock steric selections and diffusive transport[73]. It seems that too few want to remember that those electrical charges (ionic complexes, electronic clouds, functional groups), while determining short-range interactions such as protein (un)folding and electrical gradients in ion channels, are, however, in perpetual oscillation and therefore subjected primarily to long-range electromagnetic correlations[74].

The whole problem of bio-communication in the living system (be it a single cell or a complex multicellular organism) remains uncontemplable within such a perspective not only regarding thermodynamic and kinetic relations, but also and above all with regard to semantics. That is, pertinent to the fact, as mentioned earlier, that biochemical pathways and regimes are in a circular causal (and thus also teleological) relationship with respect to the biological meaning of stimuli (i.e.: with respect to 'what' a *stimulus-message-meaning* implies for the existence of the system that perceives it, i.e., that it 'structurally couples' to it)[75].

---

[72] *Chance and Necessity*, J. Monod, (Transl.), p.95, Collins, London, 1972.

[73] *The Mode of Action of Drugs on Cells*, A. J. Clark, Edward Arnold, London, UK, 1933; *Receptor Theory* by Paul Ehrlich (http://en.wikipedia.org/wiki/Receptor_theory); *The concept of matter in modern atomic theory*, M. Zuidgeest, *Acta Biotheoretica*, vol. 26, no. 1, pp. 30–38, 1977

[74] Del Giudice E., Brizhik L., op. cit; P. Renati, 2020, op. cit.; P. Renati, 2021, op. cit.; P. Renati, 2022

[75] Maturana, H.R., Varela, F.J., op. Cit.

Deferring further study of this crucial issue to another work[76], it suffices here to point out the necessary question of how a cell can see thousands of chemical reactions per second taking place inside it (and at its interface), with the participation of hundreds of molecular partners capable of meeting selectively among each other even though they are all (or almost all) co-present[77]. Regarding this, it is necessary to contemplate the key role of the super-coherent water *connectome* that allows phase correlations between the system's components in sharing their own modes of oscillation, to which correspond precise arrangements in physiology[78], which translate into reaction pathways whose outcomes, in turn, produce feedback on the very coherence from which they arose. It is then clear that the vital process is, in fact, a *thermodynamic* (and semantic) *history* along which the living autopoietic system is always the result of stimulus-response relations and is therefore the product (as it's also clear at the level of phylogeny) of a *history of adaptation* that shapes both its 'response patterns' (the process) and its form and functions (the structure).

It is therefore evident that for the recognition of a biological system as a *living being*, i.e.,, to recognise it as *life*, it is not possible to neglect the entire relational history that concerns not only its journey (during it), but also its genesis (before it). History which will provide inputs and contributions that will also determine its subsequent development (after it). As a matter of fact, life is a process of semantic relation and therefore of (bio)*communication*, which is constituted by its very unfolding, in history. It is perhaps the only condition in which *time* comes up to be represented, given that it cannot be physically given in a foundational sense[79]. Hence, infinite progressions and differences are concealed behind the single word «*life*», the quality and, if you like, the *ontological breadth* of which are proportional to the history of that system (thus also to the premises that

---

[76] Renati P., op. in preparation.

[77] *Communication and the Emergence of Collective Behavior in Living Organisms: A Quantum Approach* - Review Article, Marco Bischof, Emilio Del Giudice, Hindawi Publishing Corporation Molecular Biology International, Volume 2013, Article ID 987549, 19 pages, http://dx.doi.org/10.1155/2013/987549; *Introduction To Integrative Biophysics*, Marco Bischof, International Institute of Biophysics, Kapellener Str., 41472 Neuss, Published in: Popp, Fritz-Albert - Beloussov, Lev V. (eds.): *Integrative Biophysics*. Kluwer Academic Publishers, Dordrecht 2003, pp.1-115. ISBN 1-4020-1139-3.

[78] *Emergence of self-organization in aqueous systems and living matter,* Del Giudice E., Stefanini P., in Current Topics in Quantum Biology-9-17, Wydawnictwo Naukowe UAM, Poznan 2014, (http://www.nextcare.it/emergence.pdf); Del Giudice E., Brizhik L., op. cit.

[79] See (*Time, Cosmology and Free Will*) in italian, *Tempo, Cosmologia e Libero Arbitrio*, by Alessandro Pluchino, 2011, http://www.pluchino.it/blablabla/tempo_cosmologia_liberoarbitrio.pdf, in particular the chapters 3, 9, 10 e 11.

implied its genesis). Within such a dryly *bio-logical* perspective, clear answers are thus opened up, regarding trite and rehashed bioethical issues such as: the use of cell cloning to obtain selective developments of organs/tissues for transplantation with zero rejection, the abortion, the euthanasia, and vivisection.

What emerges is that it is not consistent to consider a simple zygote (molecularly worth of this definition because it is made up of a female gamete joined - naturally or otherwise - to a male gamete) already a "life/living being" equal to a subject /system that has a *history*, that has a *lived experience*, a *memory*, an ongoing *complexity* that is enriched at every "instant". Certainly, a seed is already a life *potentially*, and may even already be *actually* a life when it is placed in a condition to proceed and progress in its implied 'programme' of development. But a seed is not the same as a tree, even though both are "life". The thermodynamic, semantic, and relational *history* that the tree which was "only" a seed carries within itself obviously a crucial factor which has increased and enhanced that (albeit already) "life". Were it not so, we would have fallen back into the epistemological flattening whereby the right 'biochemical soup' is sufficient to have "life". This is true if, from such a term ("life"), we remove all that irreducibility that, precisely, pertains to the relational and meaning-history that implies the 'being in the world'.

Therefore, this profound distinction must be included in the criteria of judgement. If a girl/woman becomes pregnant in an unwanted way, how can she be deprived of the freedom to 'recover' her life (in the sense of the possibility of self-determination and autonomy)? Or, again, if she really does not wish to go through with the experience of an unwanted pregnancy, why must she be tainted with the guilt of a "murder of an innocent" as the religious thought often claims? This happens in such a vision because the two 'lives' are postulated to belong to the same semantic plane (which neglects the enrichment, given by becoming, by history), but this implies an unacceptable removal of a *quid* that makes them (even biologically) not identical: they are not 'worth' in the same way because they do not incorporate the same load of *sense/meaning* and *relationship*. If the opposite were true, then there would be no point in *living*, in the sense that experience would bring no enrichment (precisely thermodynamic, historical, semantic, hence "informative") to that condition and experience we call *Life*. If the opposite were true, it would have to be considered bio-logical and admissible to make test-tube babies or synthesise new creatures by gene engineering, since all what constitutes life would be that molecular, electrochemical, vibratory, etc. condition only. And we would lose everything that is not 'visible' and that concerns the semantic and thermodynamic *history* 'stored' in the path and succession of coherent states that have characterised the living system up to the present 'here and now'. But, to take again the example of the tree: while it is true that an almond is already 'a life' (ready to unfold), chewing one almond and feeding on it is not at all the same as cutting down an almond tree. The effects and their magnitudes are quite different. Just as eating an egg or killing a bird do not imply the same destructive act. And we all already know this fact because it is in biology! All we have to do is to undress ourselves of any religious or positivist beliefs and feel how the dynamics of *life* 'work'.

It is therefore interesting to note that what is argued on many religious fronts – about the fact that each zygote is in itself 'already a life' and that its use for purposes that would compromise its full development as a living being is forbidden – ultimately subscribes to the same identity collapse of the qualities of the 'life' process to mere 'molecularity', to mere 'composition biochemically suitable for biological functioning', typical of a materialist mechanicism too. Paradoxically, therefore, the epistemological flattening of 'life' is a blunder taken as much by a scientific materialism as by a theological animism, even if the latter leads to antipodal positions and ethics, both fallacious and totalitarian: the former believes that it can "play 'the little chemist' with biological soups" at will, to the extent of arrogantly placing man beyond Nature (with catastrophic consequences for the species and/or the planet); the second considers everything that is living to be 'infused' with a hypostatised essence (the soul) that connotes it *ab initio* and postulates that one has no moral right to intervene in it, especially if the purpose is suppressive (with possible dramas for the woman who cannot 'free herself' from a destiny, e.g., a parental fate, that she met despite her will) since the distinction between 'life with history' and 'life without history' is removed. The theological perspective, of course, is then plagued by the other ontological inconsistency, as will be seen in a few lines, which invokes a dualism between matter (contingent and corruptible) and the soul (eternal and necessary).

This must not to be meant as an endorsement of abortion as a desirable practice or to be advocated at all! Quite the opposite. We wish to make it clear that this choice must be freed from the same moral or existential *guiltiness* that can be associated with a 'murder' (even though abortion is an experience accompanied by foreseeable and appropriate sorrow, mourning, and due elaboration, since it is indeed an interruption of a vital process... but, let us remember, disintegrating a walnut is not the same as cutting down a walnut tree).

It is evident that the Catholic view, for example, has always contradicted itself *e fundamento* in its a priori condemnation of abortion, thought of as the murder of 'life', since it actually subscribes to that ontological flattening that isomorphises potentiality and (f)actuality. This inconsistency is the natural consequence of a (scholastic-Aristotelian) thought that has divided *being* (absolute-immutable-necessary), hypostatising it, from *becoming* (corruptible-contingent). It is precisely of the (Heraclitan) *Becoming* that *Being* is substantiated (so concerned by Parmenides and the other Eleatics and inconsistently thought of as immutable[80]). If the becoming world were an emanation of a static *Being* (be it a mathematical code, or a divine Law), how would this process unfold? That is: how can there be a relationship that co-substantiates both 'creator/cause' and 'created/effect' if the former is immutable and the latter is becoming? That is to say: how can one separate *Being* from *Becoming* while maintaining consistency and coherence, given that the creative act is intrinsically dynamic (it is a *becoming*) and implies the evolution of the 'creator'? Something is missing since one would have to contemplate a torsion point, between the static and the dynamic, which is untenable[81], unless we make an act of faith, that does nothing more than postpone and sublimate 'to a higher plane' the problem of causal relations in the structure of reality. Such a torsion point would still be the relationship between the distinct postulates of 'being' and 'becoming' and would therefore be co-substantial to both realms and thus fundamental to both. Beneath the two there is always the dynamic and evolving (becoming) *one*. The *Becoming* is right the *Being*.

What is missing, therefore, is the recognition that a *creative process* (mistakenly entrusted to a hypostatised 'god' or to a transcendent hyperuranic 'abacus' in which physical-mathematical laws exist in themselves, in truth only descriptive propositions of a language), is connoted *in primis* and foremost by the evolution of the 'creator' and that the latter is nothing differentiated from the 'created', nor from the 'creating' itself. In truth, one should stop questioning *existence* through this partitioning scheme

---

[80] (*Introduction to Parmenides*) in italian *Introduzione a Parmenide*, Antonio Capizzi, Laterza, Roma-Bari, 1995.

[81] For further analysis of this critical analysis, see Renati P., *Analogical Physical Foundations of Conscious Reality*, pp. 12-13 and the chapter *Quantization and Uncertainty*, p. 29, in which it is shown that the problem of the relationship between description and reality not only substantiates the essence of uncertainty in physics and in science in general, but is rooted in the same attitude (quantifying, quantizing) that believes a division between *being* and *becoming* possible, which is characteristic of many spiritual and religious (as well as philosophical) traditions.

because it is surreptitious and not consistent, as it is always inscribed in the artefact of the arbitrary dichotomy between observer and observed, and the diachronic distinction between *causes* and *effects* (valid only phenomenologically, at an *extensive level*) in the illusion of an idea of reality conceivable as composed of portions and categories-observable-quantities that exist in themselves.

It is necessary to transcend the logical procedure that lines up causes and effects, law of evolving and process of unfolding itself, and to begin to grasp their *analogical* co-belonging. Only in this way, thus by re-founding knowledge in the *becoming holonomy* of what that is, we can recover a choral vision of reality, in any case endowed with necessity, be accepted.

Otherwise, in this ethical "(pseudo)respectability", which eliminates the entire role of *history* in contributing to provide meaning/sense to an onto-phenomenological process such as *life*, one finds oneself in a Zeno's aporia and indirectly subscribes to a mechanistic solipsism that deludes itself in the myth of the entity/subject *per se*, as if it were already *a priori*, forgetting that it is nothing other than the result of one's own *relational history* (as much thermodynamic as semantic, therefore existential, psychic, emotional, archetypal, genealogical-systemic). And it is therefore also the history of one's *environment*.

With such an awareness, at most *in extremis*, it would then be more proper to use zygotes artificially obtained by the union of frozen gametes or by cloning cells. Such zygotes, given what it has been considered about the absence of a biological and semantic history and a 'bio-energetics of mating', cannot be considered 'living beings' *tout court* (otherwise we would be in the 'inhuman freeze' of a less than materialistic science, almost of Orwellian-novel fashion). And from such zygotes one could at most obtain the growth of specific tissues or organs for those people suffering from irrecoverable dysfunctions or excruciating pain. This, if one really, really had to, would be a wiser use of technology; and not, on the other hand, the design of semi-synthetic life, or the use of surrogate artificial wombs, which subsume an idea of life that can be fully exhausted within the biochemical-corpuscular horizon (subscribing to a fallacious dualism between structure and function, form and content, psyche and soma).

For euthanasia, finally, the same principles apply: it is true that life is not 'something' that the subject *'possesses'* and that it is a condition that determines him before he can self-experience as a subject, but it is equally true that the vital experience also includes everything that one prefers/decides to accept or not (such as an existence in a vegetative state), beyond any fictitious 'arbitrariness'.

Therefore, the requests of those who wish to escape from infernal and senseless pain[82], beyond a 'certain threshold' that each person knows and feels is valid for him/herself and which must therefore be recognised *casu per casu*, even in situations where the subject cannot choose for him/herself, should be allowed, respected and helped.

A brief comment on the equally delicate subject of vivisection or biological and clinical experimentation on animals. In this case, a historical contextualisation is due. While it is true that the study of live animal models made it possible, not only at the dawn of medicine, to understand what would happen to a human being (at least from a roughly anatomical or biochemical and metabolic point of view) as a result of certain 'stimuli' (introduction of substances, removal of tissues), it must be said that today that method/practice is now useless for two reasons. First of all, current knowledge has now exhausted the possible (albeit macabre) experiments in which one still had to understand how 'living machines' work, within which animal models could faithfully represent effects that could be extrapolated to the human model; secondly (as a consequence of the first point), what one would like to understand today from animal-model studies pertains above all to evaluating the efficacy of drugs, active ingredients or the toxic or "oncogenic" effects of molecules, or to implementing biosynthetic practices or so-called Artificial "Intelligence", which increasingly aim at distancing the human being from the web of life and at a dilution of the human-nature relationship, denatured to delusional control acted by the former over the latter.

Yet the literature is currently showing[83] us more and more how, as far as toxicological and pharmacological research is concerned, the use of

---

[82] See Malvestiti B., op. cit. ch. 2, p. 112: «it has been noted that the right to life – also enshrined in the *Nice Charter* under the I Capo – can be subordinated to the right to autonomy (Hersch 1990). This observation has relevance for bioethical issues such as euthanasia. Although life represents the *conditio sine qua non* of the possible exercise of autonomy, its preservation cannot be put before the need for autonomy, when that preservation becomes a duty: the need may lead the human being to "prefer the loss of life to the mutilation of freedom". For this reason, the question arises as to the degree of suffering or loss of consciousness from which the right to life permits the relief of death (Hersch 1990). The absolute need for autonomy is justified by the decision of the Court of Cassation in the 'Englaro case'. In its ruling of 17 October 2007, the court annulled the decree of the Court of Appeal of Milan, dated 16 December 2006, which prohibited the interruption of Eluana Englaro's artificial hydration and nutrition by means of a rhino-gastric tube. The justification is based on two principles, one of which is respect for human dignity».

[83] Just to quote a couple of references, among many: *Are animal models predictive for humans?* Shanks N, Greek R, Greek J., Philosophy, Ethics, and Humanities in Medicine: PEHM. 2009;4: 2,

animal models is useless as it is unusable for deductions on human beings, since what is being researched in such experiments concerns cyto-histological dynamics and biochemical processes that are more sophisticated and complex than those investigated in experimental 'historical precedents' and do not in any case dispense us from the obligation of trials on humans. It would therefore be right and sensible to suspend all those practices that see animals undergo pains to which one cannot be indifferent, even more so if they are not decisively useful. The case is different when behavioural studies in animal models, without practices that harm them, can be exploited to expand knowledge in neurobiology and ethology that can add value to the study of the emergence of consciousness and behaviour (including human behaviour) and the relationships between perception and physiology.

doi:10.1186/1747-5341-4-2; *Can Animal Models of Disease Reliably Inform Human Studies?* Van der Worp, H. B., Howells, D. W., Sena, E. S., Porritt, M. J., Rewell, S., O'Collins, V., & Macleod, M. R. (2010) PLoS Medicine, 7(3), e1000245. http://doi.org/10.1371/journal.pmed.1000245.

6. GMOs and pseudo-science

So, which is the point? It is in *how* we evaluate ('scientifically') the differences. As long as we limit ourselves to observing compositional, histological, anatomical, biochemical aspects, it is very likely that we will not notice anything different and, if we are blind within that criterion of examination, we will conclude that everything is fine, nothing changes («the children are strong, they enjoy excellent health!» they say about babies developed from *in vitro* made zygotes or about babies grown in uteri of women who, after birth, ceased them to other people, possibly single-parents or homosexual couples) and who cares about "the rest"? Who notices it?

As long as the epistemological substratum does not evolve to the point where 'soft' *observables* can be considered, we will continue to believe, especially in the biological sphere, that the ("objective") representation provided to us by bio-chemical-physical data corresponds to everything there is to consider. But how is it possible, for example, to claim that introducing GMOs into our ecosystems and into the diet of humans and animals is not a problem?! It is very simple: just rely on (short-term) biochemical, histological, anatomopathological and genetic analyses and consider those data as representative of what is really happening in nature.

But how many variables and correlations are we leaving aside when we think of biological material from GMOs that becomes food for another living being in which more than a thousand symbiont species (including 'viruses', bacteria and fungi) coexist and that have high gene-plasticity and are able to share nucleotide sequences with each other and with the cells of the host organism? Or how many variables and correlations are we able to take into account through *in vitro* analyses (or *in vivo* but within a laboratory or in "field trials"), compared to what happens when a new genetic code is inserted into a real ecosystem through a 'new entry GMO' in which there are tens of thousands of correlations, not only biochemical, but also and above all ethological? On what scales of space and time can the experiments, we are able to do, give us a truly representative picture of the effects? Well, the answer is: on very narrow scale ranges.

The pharmacologist and biologist (and senator) Elena Cattaneo dared to say[84] that «The GM corn that we could grow is safer for health than the others: it has fewer mycotoxins that are dangerous to humans and pregnant women and does not require insecticides that kill bees, butterflies and ladybirds, but allows these insects to live undisturbed while protecting the environment and animal biodiversity. After decades of use, there is no record of a single hospitalisation due to GMO consumption. Around the world, thousands of different varieties of GM corn and soya are cultivated (no more monocultures!) to suit different climatic conditions, increasing plant biodiversity». What kind of reasoning is that? What are we talking about? Apart from the fact that it is not true that everything is so "armless", in fact, Carlo Modonesi[85], of the University of Parma, reminds us that «The scientific literature has documented, through field and laboratory studies, a number of cases of ecological and biological risks associated with engineered crops. For example, an important comparative investigation, based on laboratory tests carried out on rodents, has shown that for three engineered maize crops (NK-603, MON-810 and MON-863) there are significant biological risks due to alterations detected in the kidneys, liver, heart, adrenal, spleen and haematopoietic organs».

But above all, it is not necessary to deal with symptoms in order to realise that something is not *in accordance* with a living system, that is to say: it is not only 'disease' in the clinical sense that we must talk about. We must talk about the fact that we irreversibly alter delicate balances and deviate the evolutionary trajectories of micro, meso and macroscopic living systems from their own 'physiological' path. Physiological because biological life is connoted by correlations and consistency constraints (both thermodynamic and in terms of eigenstates and wave functions) that make it 'move' and develop like a *holos*. If man sets out to intervene in these processes from a technical position (and instead does not 'stand in the ranks' participating in the biological dynamics on an <u>equal footing</u> with all other organisms), it happens that he can make such actions only in a manner that is already fractured and not composed to the *collective status* of the earth's *bios* and therefore the evaluations he can make, as well as the interventions it decides to make, are only made on a basis that is, in fact, mechanistic, i.e., non-synchronic, coarse, short-range and without being able to take into account the *subtle* features (i.e., about a *quid* in

---

[84] (*GMO's, let's stop the anti-scientific deception), in italian, OGM, fermiamo l'inganno anti-scientifico*", E. Cattaneo, http://ricerca.repubblica.it/repubblica/archivio/repubblica/2015/10/03/fermiamo-linganno-anti-scientifico37.html
[85] http://www.biodinamica.org/2010/09/agricoltura-industriale-colture-gm-e-biodiversita/

which, as soon as he intervenes, and participates, it is already corrupted, altered: these are the phase correlations in super-coherent systems, the biological ones[86]).

If we need 'measurements' in order to acknowledge the <u>intrinsic correlation</u> between any level, or part, of the biosphere and the rest as <u>a fact</u>, then we are falling into a glaring error of method, of the method of science (hence epistemological), which thus becomes the most acrid expression of the arrogance of a boorish logical-identitarian thought, incapable of accepting even what that has its own evidence in itself, even if it is not as tangible as a block of zinc, a chemical reagent or an electrical voltage on an oscilloscope.

What then is 'anti-scientific'? Trusting that "it will be alright" and laboratory approximations, deluding ourselves that all reality is surrogate-reducible, extractable, representable within that *toy model* (so seductive because it produces quantitative data, numbers)? Or accepting (more humbly) the fact that in Nature there are *wider*, non-trivial, and (now) non-measurable variables and *states* within an experimental setting? What is that? Did we perhaps need endocrine-immunology studies[87] to show that in developmental age the residence times (in blood and on receptors) of cytokines and glucocorticoids of non-breastfed infants are longer than those of children who instead were breastfed (with consequences on immune aspects), to know that natural breastfeeding is a biological activity of immense significance and whose effects exist even if they cannot be measured?!

Furthermore, what "biodiversity" is Elena Cattaneo talking about? We would not define as «biodiversity protection» a situation where bees and butterflies enter into a relationship with organisms totally exogenous to the ecosystem, in which they feed on pollen with engineered genetic material that could change the microbiome of their digestive systems, and who knows what else. It is rather a 'time bomb'. The biodiversity one hopes to protect is the existing biodiversity with its balance and 'tuned' relationships to the whole environment (and planet).

But is it ever possible that scientists are so incapable of giving themselves a *worldview* able of keeping in mind and contemplating this "basic

---

[86] Del Giudice E., op. cit.; *The supramolecular structure of liquid water and quantum coherent processes in biology*, A De Ninno, A. Congiu Castellano, E. Del Giudice, Journal of Physics: Conference Series 442 (2013) 012031 doi:10.1088/1742-6596/442/1/012031, DICE2012 IOP Publishing.
[87] *The Risks of Not Breastfeeding for Mothers and Infants*, Alison Stuebe, Reviews in Obstetrics & Gynaecology. 2009 Fall; 2(4): 222–231. PMCID: PMC2812877,
https://www.ncbi.nlm.nih.gov/pmc/articles/PMC2812877/

wisdom" that comes to us from an analogical relationship with Nature? Have these people ever walked in a forest in silence? Have they ever carefully observed what goes on in about twenty square centimetres of soil under the fallen leaves in a forest? Or under the surface of the water in the Mediterranean Sea? Have they ever heard the breath of a vibrating horse by putting their nose into its nostrils? Is it not the case that, perhaps, too many of those who spend their lives in the laboratory, characterising samples and trying to solve molecular structures, run the risk of having only *knowledge* of Nature, but no real *experience* of it? *Studying* a system, smashing it to pieces, does not return what the very resolving limited aspects of it destroys: what it actually *is*. The 'perversion' of bioengineering and bio-medical practices that we are denouncing is precisely and only in the *analogical removal* of that *experiencing Nature* that every human being should have as his or her first *foundation/rooting/grounding* in existence. As Richard Louv says, much of human discomfort should be recognised in what he called NDD, *Nature Deficit Disorder*[88]. There are children who have never seen a hen in real life, don't even know how humus smells among leaves, and can't get a beetle on their hand. But the Playstation's video games, the Simpsons jokes, the soccer league hits or who was nominated in the last episode of the reality show, are (useful? enriching?) things they know very well.

Well, a (pseudo-)living population in this condition is in severe danger. And in turn it is in great danger because it no longer has any *inner biological compass*, dictated by that semantic and connectional rootedness, which guides it into what is *good*, regardless of any belief, cultural or confessional orientation; *good* because *bio-logical* (logical for life). Scientists, precisely because they study Nature, should also be certain to actually *experience* it because only then they would really "Know" it. Instead, as things stand nowadays, those 'specialists', who should be able to guide us on the right choices, risk to be co-protagonists or proponents of great disasters (along with the perversion of globalised markets and oligarchic capitalism). And let us remember well that it is all Life on the planet that pays the price for missteps on this ground. Let us not delude ourselves that we can free ourselves from these inviolable roots. Only uprooting and *disease*, in the true sense of the word, would ensue.

---

[88] *Last Child in the Woods: Saving Our Children from Nature-Deficit Disorder*, Richard Louv, Algonquin Books, 2008

So let us all wake up from reckless, unconscious, "science-techne" and let's begin to build synergistic a true *Co-Science* based on presence and listening.

7. *Not* learning from mistakes

Before moving on to our final reflections, let us dwell again on the GMO issue, since it is very representative of the same problem raised by the creation of semi-synthetic life through the expansion from four to six bases of the genetic alphabet. Let us first point out why the immediate benefits arising from their use must not, seductively, replace or weaken our serious commitment to solving problems on a more upstream level. It is time to stop dealing only with the symptoms.

The absurd danger of pursuing a research project aimed at producing semi-synthetic biological organisms should not, at least after these initial reflections, be difficult to grasp. So, GMOs, cis-genics and any manipulative action at the level of, let us say, biological "software" fall into the domain of problematic practices. There are already historical precedents that should have taught us the lesson. Just think of what happened in Australia when, in the mid-19th century, 24 rabbits (12 couples) were introduced by Thomas Austin in order to have some game to hunt (for personal enjoyment!): the rodents soon multiplied exponentially and quickly caused disruption to the soil due to the immensity of burrows dug for making dens and escape-routes, and the survival of other plant species was endangered by the enormous biological pressure exerted by the rabbits, which, in an environment devoid of predators, increased disproportionately. Moreover, other native species (marsupials) were threatened by this expansion as they found themselves with food and space resources decimated by the competitive role of the species that (there without natural predators) had quite different biorhythms in terms of nutritional requirements, metabolism and reproductive rate. The first "brilliant" move to remedy the disaster was to introduce foxes (well-known rabbit hunters in European or North American ecosystems) to reduce their demography. Result: in the presence of many other palatable preys, slower and easier to catch, we found ourselves on the brink of a catastrophe with several native marsupial species at risk of disappearing and with foxes and rabbits in rampant expansion. The containment and mass culling of foxes and rabbits had to be attempted. But of the latter, the numbers were now disproportionate, so, again in a naive (or unconscious) view, other 'biological' strategies were adopted. Here is how Wikipedia relates the rest of the story[89] . «Until the second half of the 20th century, government counter-measures were ineffective. Only in 1950, with the spread of a mosquito vector of a powerful "virus" [maybe better to say: a toxin, a.n.], Myxomatosis, the [rabbit] population was reduced by more than 80%, resulting in the death of about 500 million individuals. Among the surviving individuals, however, a resistance to the disease was developed and they began to breed again. In the 1980's, the spread of Myxomatosis-resistant rabbits led to the resurgence of the overpopulation problem. For this reason, another "virus", RHD (Rabbit Haemorrhagic Disease), was introduced, which led to the death of 90% of the rabbits, but again caused the emergence of resistance strains to the disease. The latest 'weapon' against this scourge is a new form of Calicivirus, RHDV (gen. Lagovirus), introduced in 2000, which has caused the death of so many rabbits to avoid more than $1 billion in damage in 10 years». However, today, the annual cost to the Australian economy due to rabbit management is around $113 million because rabbits feed on vegetables, fodder for breeding herbivores and young seedlings/sprouts of shrubs and trees, effectively rendering any reforestation work useless. Moreover, far more serious (than the super-structural monetary damage), there is the great risk that the 'viruses' released into the ecosystem to counteract them will have a no less serious impact in the future, as the virological weapon used (the RHD) would appear to be lethal, not only for rabbits, but also for other animal species found on Australian soil, thus potentially causing new anomalies to the delicate natural balance. By now, the only thing that can be done is to hope in the processes of compensation, adaptation and intelligent self-regulation of ecosystems, understood as *living subjects* capable of moving as a *whole*.

This example, regarding the reckless introduction of an allochthonous species into an ecosystem, should teach us that in ecology one cannot adopt linear thinking, since everything depends on everything and the *properties emerging* from any perturbation are unpredictable in an integral sense. One can at most make estimates, but nothing better. And estimates are not enough when the possible price to be paid is so high.

Again, Cattaneo wrote: «While the importation of 50 different types of GM plants was authorised without saying anything, the cultivation of GM corn, judged on scientific grounds to be safe for the environment and health, was prevented. In Italy (non-GMO) corn is attacked by pests and its yield has dropped. In 2004 we were self-sufficient, today we import

---

[89] Italian page: https://it.wikipedia.org/wiki/Conigli_in_Australia

almost 60% of it (also GM corn). Billions wasted. This is 'politics' that decides. On the fact that our animals are fed with GM feed bought abroad (85% of soya, almost all of it GM) everyone is silent. The current government could have adopted a new European clause to stop importing GM feed. But no. On the contrary, it falsely claims through a minister that the country is "GM-free", with the same hypocrisy as the marketing of the multinationals, who swear to consumers that they are 'against GMOs' and keep silent about the fact that the cheeses and hams they sell come from animals fed 'full GMO'. Information that is missing on product labels. And 'politics' looks at somewhere else. If there were transparency, all our high-quality products (those of the great protection Consortia with which we wean our children) would have to be labelled as "derived from animals fed GMOs"» Well, we could of course agree with such a reporting, but which is the conclusion that Cattaneo draws from all this, even correct, criticism and denunciation? This: «For 15 years, legislators have assumed that GMOs are dangerous to health, the environment or biodiversity. Yet, studying one by one the interesting ones, this is not the case. The GM corn we could cultivate is safer for health than the others: it has fewer dangerous mycotoxins...» etc. (by continuing with what we have already reported above). Unbelievable. What?! Due to the fact that political-economic agreements have caused that GMOs 'stopped at the door' could sneak in "through the window", and that this costs Italy dearly in economic terms, the proposed solution is…. "Let's produce GMOs ourselves, so we save money and use less pesticides". It's absurd: instead of concluding that they are awful stuff that should not be fed to either humans or animals and that they should not even be created, as they produce potentially unmanageable effects, she promotes them as both an economic and biological solution!

Honestly, we wonder how molecular biologists, pharmacologists, bioengineers, geneticists (and every thinking human being) cannot be dismayed at the idea of transferring the same recklessness – just seen on a macroscopic scale in the 'rabbits in Australia' case – down to a sub-cellular and molecular level. Once a biochemical "upheaval" has been introduced that becomes part of the regulatory "software" of biological autopoiesis (whether in the case of the addition/replacement of genes as in GMOs, or for the far more radical shift due to the change of 'alphabet' given by the introduction of two new genetic 'letters' as the Romesberg group was doing) who can do something to remedy it anymore?? We are faced with a multiplicity of scenarios that are indeterminable *a priori*, not a few of which involve potentially catastrophic outcomes. In the sense that 'apoptotic' processes can occur at the level of ecosystems in which various species and/or *biological niches* disappear due to changed behaviour, changed feeding habits, the breakdown of symbiotic processes, new production of molecules that do not leave the rest of the *bios* indifferent, etc. And this can happen in 10 years as well as in 100; for natural evolution, these are always 'non-physiological' modes. And, even in the case of faster kinetics, these are changes beyond the scope of monitoring feasible through our laboratory studies, especially for the spatial range and number of eco-components they would involve.

But you don't need an IQ like Kurt Gödel or Werner Heisenberg to understand this! Is it possible that there are people in the world who do not feel like elephants in a crystalware shop at the mere thought of acting on Nature in this way? This comment is intended to ignite in those who have not yet dealt with the question precisely this kind of caution, sensibility and humility!

8. Agricultural technocracy and zootechnical genocide

The defenders of GMOs, on the other hand, convincingly bring positive examples of how the use of these new engineered species avoids disruptions on other fronts, such as the use of pesticides in the cultivation of certain plant species in agriculture (from cereals and soya for human consumption and livestock, to poplars for the paper industry and species for gardening). Another case, then, in which the pros of GMOs are flaunted, is the possibility of the production of propane gas through biosynthesis (even under aerobic conditions) by genetically modified strains of E. coli (always them!)[90].

Here too, just as when health care and therapeutic weapons derivable from 'playing' with genes are proposed as 'promising' and valid motivations, it must be carefully observed that solutions are still being proposed subscribing to the same *setting* that gave rise to the problems in origin! And these paths are chosen because they are technically feasible and expendable within that already existing techno-economic system. Much more difficult, on the other hand, seems to be to change the approach and mentality by which ([in]civilised) human beings relate to their ecological environment and their human fellows. Let us explain this better by starting with the second problem and then returning to agriculture.

Replacing fossil fuels with other fuels is certainly not the best option since 1) these are ways of producing energy (with combustion) that still hold the same critical issues such as the however questionable unproven "CO_2 increase" and the other already (over)known problems related to the availability and distribution of such burnable resources; 2) there are other technological prospects which, if duly incentivised, could provide us with that much-needed coverage of our energy needs (from renewables to other frontier chemical-physical processes − such *as Low Energy Nuclear Reactions* or the exploitation of '*Brown's gas*' from the electrolysis of water, or the exploitation of the motion of permanent magnets to produce exploitable excesses of electrical energy with respect that required for their motion − which are still too marginalised by the mainstream science because they are considered 'impossible', yet something is moving).

---

[90] http://www.lescienze.it/news/2014/09/04/news/batteri_produzione_rinnovabile_propano-2272252/

Returning to agriculture: if even it really happened, as the promoters of GMOs claim, that the cultivated non-GMO plant species today contain mycotoxins (such as cereals) or if they require a lot of pesticides to give successful harvests (cereals and legumes), it is because irresponsible farming is practised, based on models that are incompatible with biological functioning.

Industrial agriculture has already been shown[91] to be a losing and unsustainable choice, profitably replaced by synergistic or permaculture techniques in which different species (both plant and soil symbionts, such as nitrogen-fixing mycorrhizae, ethylene-oxygen cycle involved bacteria, fungi, insects, and various invertebrates) constitute a *whole* to form an ecologically stable and synergistic niche, i.e., one in which the various species benefit each other. And this can very well be implemented even on a large scale and does not necessarily exclude the possibility of using machinery that facilitates certain operations: alternating various species in rows, for example, allows for the use of means in the preparation of the soil beds, in any light weeding, mulching, possible irrigation (preferably drop by drop) and for harvesting.

Another serious aspect of industrial agriculture has been the ousting of many native species of fruit, vegetables and cereals in the various territories, favouring hybrids created *ex novo* (almost always sterile, so that the seeds have to be bought again each time!) or already existing varieties that guarantee higher production/cost ratios, longer *shelf-life*, greater resistance to parasites, vegetative forms more suitable for automated processes such as drupe harvesting, threshing, etc. But all these are choice criteria that apply when one subscribes to a production setting such as the standardised one we are talking about. By changing the way we farm and relate to the land, there is no longer any need to use those criteria to choose what is profitable to cultivate or not, and we can return to an original biodiversity in which those who live in a place produce the species most suited to that place because they developed phylogenetically there, and capillary production is promoted in agriculture to minimise transport and prioritise the health of what is cultivated and those who benefit from it.

Let us take a couple of examples: corn and soya. Corn is a drought-resistant species in origin; the Olmecs and Mayas cultivated it in Central

---

[91] *The Fatal Harvest Reader: The Tragedy of Industrial Agriculture*, Andrew Kimbrell, ISLAND PRESS, 2002, google books link:
https://books.google.it/books?hl=it&lr=&id=plTcVDph_SQC&oi=fnd&pg=PR5&dq=unsustainable+industrial+agriculture&ots=_0zTvFA9S4&sig=FBBe-ucsX-YgGcNWE36woLuMKf4Q#v=onepage&q=unsustainable%20industrial%20agriculture&f=false.

and South America for centuries in torrid climates and in soils that were anything but wet. In Europe and America, starting after the war, with a goal to having cobs with more grains (caryopses), larger and with a greater exploitable mass on the total harvest, they selected and/or hybridised it to the point of turning it into an «aquatic monster» that - as Dominique Guillet[92], founder of Kokopelli, an association that produces and distributes ancient natural seeds, well explains – «guzzles 120 litres of water for every small plot of land[93]». Today, water supply problems have arisen because the aquifers are increasingly empty, or the general demand (over civil, industrial, livestock and agricultural users) has increased dramatically. Thus, the seed and agrotechnical industry proposed GMO varieties, saying that they were developed because they were "drought-resistant"! Result: in origin we had species that already resisted drought very well, then they produced varieties that swallow water, then – when the water supply ran out or the supply was reduced – they ended up with transgenic corn because – they said – we need species that resist drought! Instead of having preserved the old maize varieties, which are naturally suited to water scarcity, all the prerequisites for viable agriculture have been alienated, and the farmer's dependence on the industrial and biotechnological system inevitably has been sealed up. Is all this honest and truly profitable?

The hybrids or GMO species, moreover, contrary to what is often boasted, are plants precisely *for* pesticides, in the sense that they are either organisms that produce them by themselves (by metabolic means, implementing a competitive biochemical competition with respect to pest species) or they are designed to resist the pesticides that are dispensed on the land. The best-known case is that of Monsanto's patented Roundup Ready® soya, capable of resisting the very powerful glyphosate-based Roundup® herbicide produced by the same multinational company. Dispensing the Roundup® to eliminate the weeds in a field of 'normal' soya would mean its immediate death, whereas transgenic soya is resistant. Imagine, however, what it might mean to eat (for humans and animals) a legume that during its biological cycle has been in chemical contact with (and possibly absorbed) glyphosate. Not to mention the biological damage to the micro flora-fauna in the soil, for wastewater from the crop and the

---

[92] See at time 1h:27m of (*Local Solutions for a Global Disorder*) in French, *Solutions locales pour un désordre global*, documentary, directed and screenplay by Coline Serreau, 2010.

[93] Better to specify: considering an average density of 7 plants/m², and knowing that the average water consumption for irrigating corn today is about 6000 m³/ha (see for example www.assomais.it), this means to get about 70 thousands plants per hectare and a water consumption of about 85 litres per plant!

dispersion into the air. Worthy of mention, again from Serreau's master-piece, are the reflections of two well-involved and knowledgeable women on these issues. The first is Vandana Shiva (a nuclear physicist who later took up the ecological cause), who reminds us: «we have to see the difference between good food and bad food. The market does not make this distinction. If we fail in this, not only we will no longer have freedom, but we will also no longer have bread. This link between the field and the table, which allows the good food produced by farms to reach every kitchen, is the reinvention of democracy. Because if we do not maintain this link, we will no longer know what we are eating and we will be forced to eat what human beings should not eat, what no intelligent species should ever eat». The second is Ana Primavesi (Brazilian agronomist), who reports: «GMOs are just an adaptation of crops to dead soil. So, I ask: is it really necessary to work with dead soil? No».

Indeed, let us deepen this last very important point. Another mythological devilry is ploughing the land. At most, if it is necessary, it can be carried out the first time one prepares wild soil for agricultural use; after that it becomes a guarantee that that soil will always be dead and incapable of developing active micro-sites for the unavoidable symbiosis between micro-organisms and plant roots as it mechanically destroys them. Consider putting the lung tissue in a blender and then see if it still works as an oxygen-carbon dioxide exchanger (the chemical composition is the same as when it was intact, right?!).

By now it should be more than obvious that the maintenance of the life process lays on micro-compartmentation. To keep entropy low within each work cycle (at any scale, from the sub-cellular scale of biochemical cycles to the water cycle between the atmosphere and the earth's crust), biological material compartmentalises into hierarchies of systems fractally made up of other sub-systems and so on... and the same happens at the macro scales, in ecosystems[94]. This allows work-cycles to take place in which the entropy produced within that compartmented structure is practically zero because it is dissipated externally, in the hierarchically just successive (and spatially larger) system which, in turn, will do the same and so on... up to the biospheric level. Let us look at a cell or even just a mitochondrion and immediately understand how compartmentation

---

[94] We advise to well study the work of Mae-Wan Ho, *The Thermodynamics of Organisms* (http://www.i-sis.org.uk/ThermodynamicsOfOrganisms.php) and the thermodynamics of dissipative processes by Ilya Prigogine (*Introduction to Thermodynamics of Irreversible Processes*; Prigogine, I., C.C. Thomas, Springfield,'54; *From Being to Becoming*; Prigogine, I., Ed. W.H. Freeman and Co. S. Francisco, CA, USA, 1980).

(e.g., by membranes) and entropy zeroing, coupled with the maintenance of coherent long-range excited states, are the basic requirements for any biochemical activity carried out, since molecular reactants do not come together according to diffusive processes, but rather in precise order within time-dependent and autopoietic work cycles[95] (such as Calvin's and Krebs').

Well, in the soil, as a biological niche, the same condition in which precise micro-sites and topological structures can exist is necessary for it to be alive and therefore to dissipate entropy. In such a condition any kind of manure or fertilisation is superfluous (when not counterproductive as it can trigger fermentative processes that would compete with the reaction cycles – mainly oxidoreductive – involved in inter-biological synergies). In fact, plant species do not build mass through the amount of organic matter present in the soil, but through air respiration and water. What they need are *living mediators* to fix all those substances that play a key (or catalytic) role in their metabolism (from nitrogen to ethylene, to trace elements, etc.)[96].

Again: how is it possible that geneticists, biologists, and agronomists are not all in agreement and solidarity on supporting and proposing this type of solution to companies, governments, and the people through an honest scientific disclosure? Recalling again the words by Carlo Modonesi of the University of Parma, for the simple fact that «what we can and must say today with all honesty is, after all, even more worrying. No scholar in the world, in fact, has the slightest idea of the extent of the effects that genetically modified plants will have on a global ecosystem scale», there should not even be any doubt about whether or not we should allow ourselves "to play Lego®" with Nature!

To tell the truth, still talking about solutions on a level other than the one within which the problems have been produced, the important contestation to be made is not the one that brags about whether production flows are sufficient, or not, to meet the demand for raw materials for food and livestock, but rather the one that questions precisely the average eating habits of modern societies (in both East and West) regarding, especially, cereals and meat. Who benefits from eating huge quantities of products

---

[95] *Water Dynamics at the Root of Metamorphosis in Living Organisms* E. Del Giudice, P. R. Spinetti, A. Tedeschi, *Water* 2010, 2, 566-586; doi:10.3390/w2030566

[96] We should take a close look at Masanobu Fukuoka's studies (*The one-straw revolution*, Rodale Editor, 1980, Original title: *Shizen noho wara ippon no kakumei*, Hakujuso Editor, 1975) and by Emilia Hazelip (*The Art of Cultivating by Letting the Soil do*, in Italian *L'Arte di Coltivare Lasciando Fare alla Terra*, http://www.silentevolution.net/docs/Dispensa-Agricoltura-Sinergica.pdf).

derived from flour (which is, moreover, mainly refined, and impoverished) and above all from eating meat from animals *devitalised* by asphyx-
iated existences, encapsulated in productive niches and whose tragic
deaths would horrify the most impassive guy among the insensitive ones?
Paradoxically, if the immense spaces subtracted from massive cultivation
for intensive livestock farming were given back to the wild to become
woods, forests, heaths, clearings, and we hunted (maybe with agreed
plans) our game ourselves, the ecological impact would be far less, both
in terms of pollution and in terms of biodiversity and environmental con-
servation. And the meat of animals that have truly lived and experienced
the relationship with their natural surroundings, expressing their full existences, would be of a far different quality. But, of course, this makes the
'sober scientists' laugh, since if the 'chemical composition' is the same,
then there is no objective difference... They can continue to believe this.

Are we proposing a naive return to "the wild"? We wouldn't say so,
rather a move towards the *logical-bio*, that original condition that recognises the human being's right to explore and to be free to know his environment (and what he feeds on) through a corporal, intimate relationship,
made up of searching, gathering, predation and satisfaction from this expression and relationship.

Only those who have lost the intimacy of the body-relationship with
earth, leaves, scales, blood and breath can believe that hunting or fishing
are now signs of incivility. In the intimate relationship and confrontation
with the wild, there is the knowledge of an authentic world, which is ours
and which gives us meaning; in the prophylactic distance of an intensive
farm or a 20-kilometre trammel only the cowardice of a hypocritical horror concealed in vain.

It should be noted that even the "packaging" of natural spaces within
'reserves', 'parks', and off-limits areas in general, where the only activity
allowed is often that of the insipid 'visitor' or 'walker' is a sad expression
of the multiple impoverishments of contemporary consciences.

First and foremost, it is the subscribing to a cognitive modality that
has prescribed an exclusively visual, eidetic relationship with reality,
which is bi-dimensionalised into an image, a *spectacular scenario* of
which to be only spectators, virtual and external users, as in a movie.
Therefore, this also expresses the sad deprivation of the human being –
who seeks a genuine and sacred possibility of re-soaking in that ancestral
root that inhabits us – of the rediscovery of our own animal motives of
predators, explorers, seekers. All motives that rediscover an analogical
perception of the environment and a knowledge of it made of *feeling*. And

therefore of great *care*, exactly the one which, if lacking, then requires ultimate actions (like prohibition to do…), which, however, impoverishes everyone enormously on the fronts we are listing.

Indeed, if one can justify the creation of 'protected areas' *et similia* as a *last resort* to safeguard from insane gestures (individual such as poaching or collective such as urbanisation) the now just traces of an ecological universe with which we would be in perfect harmony and symbiosis, such a practice only testifies to the fact that we have betrayed that universe of ours through the technological *hybris*, which in fact prevents an equal relationship with our environment and other species.

Finally, such 'confiscation of Nature's beauty' is proof that even the space to live, to know, to contemplate, to venture, to find one's own corresponding dimension (especially if characterised by peculiar charm and splendour) becomes *de facto* private (i.e., deprived, taken away) property. And the holder-owner of this good, which – like air, water and every vital resource – should be without any patronage, is a non-subject (a no one): the 'system', the same that founds its false necessity and existence on those same categories, structures and criteria that see in the forms/categories of the *norm*, the *commodity*, the *capital* and the *spectacle* their inviolable totems; mere consequences of the human substratum now devoted to *malice* (Nietzsche) and *measure* (Musil) by which the meshes of the anthropological *project* (Heidegger) are woven.

9. Re-mediating the techne:
beyond eidos, bits and quanta

The difference between an indigenous and a non-indigenous person lies in the kind of connection with their places. Indigenous peoples think that the earth and sky encompass them, that they (earth and sky) have rights over them, that they own them. In reverse, non-indigenous peoples believe that they own the land, own the water, own the sky. An indigenous people is made up of humans who believe they belong to a place; a non-indigenous people is made up of people who believe places belong to them.

Derek Rasmussen, *L'Ecologiste*, no.17, Spring 2006, vol.6, no.3, p.28.

Well aware that the many themes sat on these pages would each deserve far greater investigation, the aim of this text is indeed to trace a coherent thread that connects these themes from the side of their ideological and epistemological premises and that allows for their critical observation through an awakened sensibility.

In fact, the focus of this paper has been to highlight how human history is increasingly condensing into a 'deadly binomial' that sees *technology/techne* and *deliberate will* in a self-propelled cycle in which the former originates from the latter, without any mediation provided by the consciousness of being (humans) as *relational processes*, and in which the latter exploits the former to fulfil itself unconditionally, in the illusion of an individualistic identity (of the individual or of the human species) that must prevail and go ahead along 'its own way' at any cost. As if it were isolable or above, beyond, the ecosystemic *womb* or could be less dependent on it than other living forms.

In all the biotechnological practices discussed in the previous chapters (from surrogate wombs to heterologous fertilisation, from the creation of organisms with genomes with six nitrogenous bases to GMOs, from genetic editing to reckless agri-zoo-technological development models), *techne* (in itself even possibly positive and redeeming) becomes both the *trigger* of the delirium that underlies them and the *guarantor* of their implementation, i.e., the instrument that translates such delirium into factual reality. This deadly binomial is today made even devious, almost without return, as the human being inhabits and feeds a technical environment that today has become above all a *medial* environment, in which that is to say, the *media* (the technological instruments of communication) are no longer 'tools' to be used, but real *scenarios* of social action and bionic *prostheses* that we wear and that have rewritten our aesthetic subjectivity. This media environment is therefore a *perceptual* environment in a dual way: in a *functional* sense as it moulds our perception and changes its modalities (think of the concept of the optical unconscious introduced by Walter Benjamin[97]) and in a *semantic* sense as it modifies the internal *aesthetical* scenario regarding tastes, emotions, imagination and, above all, *desire*.

The pervasiveness of the medial environment causes an atrophy of the subjective aesthetic, collapsing the subject into a homologising collective horizon. That same horizon that makes people believe they want something (such as goods, social status, children, animals, prostheses...) because it silences a deep and hidden vocational emergence, but crucial for anyone to be able to acknowledge oneself as such (i.e., *me*).

In this silencing, the horizon of incessant communication and the spectacularization of any expression submissible to the *commodity* category, in which 'civilised' human beings are almost drowned, play a decisive role. *Techne* becomes sponsor of itself and, by being bought as an instrument to fulfil the possibilities it has envisaged, in fact buys all freedom of *being subjectively*. In *The Society of the Spectacle*[98], G. Debord expresses very well how in the development of the *media* (as early as the 1960s) the *spectacle* constitutes «the autonomous movement of the non-living», «death disguised as life», something that is not properly real, but which gives us the illusion of being so. It constitutes and guarantees the construction of a *regime of desire* through the power of images, which *project* us, designs our lives as mass-consuming spectators, which designs our passivity. The «*solitary crowds*» appear in this *regime of the image* that implies emptied relationships that make us lose real touch with others and that do not allow us critical distance. Pseudo-relationships that constitute an impersonal sphere of tastes, feelings, emotions, desires that involve us entirely and that do not allow a process of internalisation of any stimulus, but only its phagocytosis.

This is how, for example, a homosexual subject no longer accepts his biological nature with dignity and self-love, but seeks to make himself

---

[97] In *The Artwork in the Age of its Technical Reproducibility*, 1936, W. Benjamin speaks of *optical unconscious*, with particular reference to cinema and photography, as that mode of vision and imagination (image creation) that is not innate in the visual capacities of the human eye, but technologically induced through the techniques that can be implemented thanks to the camera (freeze frame, slow motion, sequence frames, zoom, etc.). The unconscious imaginative mechanisms are changed through the experience of a technically mediated, and technologically made possible and experienceable, vision.

[98] (*The Society of the Spectacle*) *La società dello spettacolo*, Guy Debord, 1967.

"equal" to those who can... have children; this is how many mothers must-want to work because they are people who "can do and manage on their own too" and prove themselves autonomous... instead of questioning the current socio-economic system that, instead, by default, should guarantee all the means for a dignified life to every mother or housewife and without making her dependent on the earnings of third parties (husbands, sons/daughters or relatives in general), and recognising that infinite, invaluable vital and invisible "work" that multitudes of women, mothers, housewives do in caring for their family, in listening to their men and children when they return from their "hard days", in always providing them with an environment worthy of being called "home", and much, much more... ; this is how corn, which must be produced (in a way that is not at all questioned, but which in truth constitutes the problem) must be genetically 'updated' in order to obtain hybrids that allow that agri-techno-economic merry-go-round to take place, with the presumption that Nature is to be "corrected" in the face of techno-capitalist distortions; this is how one thinks of creating genomes with a new (implemented) genetic alphabet; this is how one thinks of "improving" the human genome with editing practices, etc. . . It is the festival of *deliberate will*. Technically one can, therefore, one does.

And so, in the present age of the «integrated spectacle» (again with Debord), "the perfect crime"[99] is enacted and the *spectacular horizon* becomes the opposite of *dialogue* and *listening*. Through the eidetic force of images and the *devices* that transmit them, one loses his/her own active role in perception (primarily of him/herself) and the aesthetic attention becomes as ubiquitous as it is anonymous. In other words, we are faced with a social scenario today that is indeed *aesthetic*, but impersonal and projecting, programming, since feeling is no longer 'my own', but is a *feeling* already given, a feeling already produced *elsewhere* that only wants to repeat itself and be repeated. The pervasiveness of a feeling produced by the *media*-cracy entails the construction of a powerful *seduction* that collectively orients our desire, and from which it is very difficult to de-situate, which it is almost impossible to resist.

This is why *techne* – when, if, not mediated by that sensibility proper to the *indigenous*, inhabitant of a planetary ecosystem – becomes the keeper of the mortiferous and denier of the "speech of life" (the bio-*logos*). The first step lays in the fact that technology presents and materialises a-biological scenarios in which one can do (technically) something that was

---

[99] Debord G., op. cit., in the preface by C. Freccero and D. Strumia.

not contemplated in the natural horizon of things, nourishing a desire in anyone who is identified with an artificial, superstructural and projective polarisation of him/herself and at the same time de-subjectivised from his/her own biological and archetypal *self*. By way of example, think of those who believe they are performing a gesture of generous altruism in 'lending' their womb to give birth to a child who will be torn from that same womb; think of those who believe they are "increasing" the genome's coding power by augmenting the number of nucleotides, deluding themselves that a biological system is isomorphic to a digital process; and think of all that has been discussed in the preceding pages. The second step, then, lies in the fact that, when the possibility of *being able to do* things technically presents itself with all its bewitching power, such an alienated humanity, already oblivious to the logos of nature and the wisdom of the body, falls victim to that (narcissistic) seduction that promises an existential redemption and a self-realisation that in truth turn out to be the greatest failure to a sincere *individuation* (with Jung) and of a concrete *formation* (*Bildung*, with Adorno), sanctioning only the triumph of that homologation, *designed* through the pervasiveness of the spectacular *media*-cracy and through the "collectivised by default".

Resisting the *medial* and *technological seduction* is only possible through a path of personal *training* that is primarily constituted by bodily and analogical experience of the environment, of others and of ourselves; one must listen deeply to the body to understand its wisdom and feel how much and what it really needs (bio-logically and semantically) to be said *to be alive*. This alone allows access to a *knowledge* that is not limited to the 'grasped information', to the 'provided data', to the 'learnt notion', to the 'memorised scene', to the concept one reads, to the 'sponsored product', to the 'acquired emotional model'... and much more.

And in this 'much more' there is first and foremost the epistemological paradigm of reference, the one within which scientific thinking is unfolded, the one that posits *how* one searches in nature, *how* one recognises the *true*. The removal of the analogical aspects, in the form of a minimised bodily experience of the relationship with nature and with others (sometimes even replaced by squalid or insignificant promiscuities, as much social as subjective) results in the structuring of a *forma mentis*, likewise, incapable of seeing the real in a relational manner and more readily believing in the myth of decomposition into parts and of the representability of a *whole* that is in truth not epistemologically reducible to a sum of enumerable pieces.

In *The Post-Modern Condition: A Report on Knowledge*[100] (1979), Jean-François Lyotard well highlights a fine nuance of the dangers produced by the unrestrained enhancement of communication through technology, which today is almost synonymous with 'information technology'. Lyotard points out how the development of 'informatics' and the processes of digitalisation of messages and of information has resulted in the progressive depletion of the fundamental model of the *Bildung*, i.e., the *formation* of man in the West. Adorno defines *Bildung* as *culture on the side of subjective re-appropriation*, contrasting it with pseudo-formation (*Halbbildung*), the definitive death of all culture, meaning by 'pseudo-formed' the individual *user of an aborted culture, socialised*, and made immediately accessible without preconditions[101]. Precisely what often happens on the internet.

As the informatization of knowledge proceeds, in fact, what increasingly permeates knowledge and what is increasingly accepted of it is that knowledge that has been translated into informatic/digital language, which will circulate socially in such a way as to transcend the relationship between teacher and student/trainee/learner (between the one who knows and the one who learns), which is a relationship that corroborates the growth of the person (both the pupil and the master). Knowledge is accessed from the agencies that have translated it into digital data, computerising it, and it is acquired (and 'consumed') like any commodity. Bearing in mind how essential the person's *Bildung* project is for the creation of a gnoseological, intellectual and social substratum capable of critical perception and mediation of human action, we cannot ignore the pervasiveness of the *medial* environment and of new technologies. In fact, it is a matter of a spiritual, and not only cultural, formation of the subject, resulting from any form of education relationship (as teacher-student) and learning, even autonomous, provided it is in the name of critical thinking.

The new *media* (like internet in its various forms of chat, blogs, posts, social networks, browsers, search algorithms...) re-*mediate* (mediate again) the previous media (television, radio, entertainment, newspapers, propaganda...) and renew and accentuate the dangers to the freedom of the subject that the old media had already produced and that have been discussed above. It suffices to think of television and the epidemic incisiveness on mass communication processes over the social sphere, and to think

---

[100] *The Postmodern Condition: A Report on Knowledge*, J-F. Lyotard, 1979.
[101] Cf. Adorno T.W., op. cit.

of the extent to which they have shaped the collective imagination and the regime of desire.

Within today's *medial scenario*, then, knowledge, especially scientific knowledge, is increasingly actualised through the categories of number, quantity, the discrete, the bit, the quantum, and it is right in this instrumental substratum of theory the technical delirium takes place where research is made to express its own power or to unveil other versions of exploitability within the assortment of what is known.

It is in the midst of this landscape that the *hybris* which removes the difference between (i) the extraordinary and extremely useful potentialities offered to us by digital machines, by the processes of *discretization* of signals and of the (physical and logical) observables and of their translation into a form that can be submitted to calculation, expendable, and (ii) the infinite *continuum* that underlies the possibility of living systems (humans) to be creators of such processes, but much more than those. The diffusion in the common sense of this removal reaches its condensations in paradoxical expressions of speech and research projects, such as: DNA as a "digital memory" on which data can be stored[102], the brain thought of as a "computer" and neurons as transistors, the drive to create biotronic machines and to implement more and more biological systems with informatic/digital ones. This also happens under a "prosthetic guise", without knowing that, while producing very useful applications such as, for example, the recovery of motor and sensory functions of impaired or injured subjects, in this conjunction of what is artificial/synthetic and what is biological, the resultant functioning is downgraded to the rank of the possibilities of the least complex and holistic system: the artificial one. This is a poverty that is not only functional, but above all dynamic, i.e., the capacity to respond, relate and adapt in time according to variables that cannot be measured, since they are non-local and synchronic[103]. Therefore, it is an excellent practice when aimed at recovery from damage, it becomes perverse when understood as a deliberate bionic grafting procedure.

Being the creator-discoverer of the digital realm, in fact, places the biological human subject on a functional meta-level with respect to what he has created, testifying to the living system's irreducibility to a mere 'sum of parts' and the inadequacy of any conception of it which subsumes

---

[102] https://www.wired.it/scienza/biotech/2017/07/13/film-dna-memorizzato-batterio/?utm_source=wired&utm_medium=NL&utm_campaign=daily
[103] Stefanini Patrizia, Del Giudice Emilio, ., *Emergence of self-organization in aqueous systems and living matter*, 2014; Renati P. 2015, op. cit.

a mechanistic, diachronic, causal-linear (for *bits*) or probabilistic (for *q-bits*) functioning.

In an informatised and cybernetic society, the *digital* and the quantum-like (which have in common the category of the *discrete* as a fundamental quality of what is described) no longer become mere functional elements and instruments of a circumscribed[104] operational sector, but rather become a total existential horizon, almost finalistic, a quintessence towards which to strive integrally and into which to lead everything that can be experienced. The entire scenario of human action, not only technological, but right up to the relationship with biology in its ever more pressing synthetic gemmations, has as its submerged motive the tension towards digitalisation, the discretisation of reality and its irreversible (fake) laceration into *software* and *hardware*, in a perfect "mimesis" of an electronic calculator. All this reveals the implicit and ubiquitous tension towards reproducibility, storage, re-writability and the consequent decoupling of the *message* from the *medium* and of the entity from its context/environment. As we can see, Descartes and Plato, with regard to the irremediable ontological duality they formulated, are die-hards.

This is why and where *media* informatization becomes dangerous: because, as Lyotard already grasped, it prepares a cultural terrain and learning environment in which categories, criteria, tools, and sensibilities are already fenced within the barricade of the *discrete*, of the monadic and *identitarian*. In this way, all the *analogical* (features) is lost, all that *feeling* of reality *first* as a *hólos* and, if anything, *only then* knowable, describable, through analysis and decomposition, in the full awareness that these practices will always only be able to give us approximations of what really is[105].

---

[104] Circumscribed because it concerns a grammar that is then translated into multiple applicative extrinsications - the *bit* - or circumscribed because of the premises underlying the formulation of a theoretical framework with which to view and describe the world - the *quanta*.

[105] Renati P., 2016, op. cit.

Conclusions: an appeal

We need to start looking at nature with systemic eyes and not "in detail" (not primarily). It is necessary to take ourselves to a sufficiently refined 'level' such that every human being, scientists and researchers first and foremost, really knows how to perceive him/herself essentially and fundamentally as a *relational* process within a biological *logos* that transcends the individual represented subject, while still connoting it in its immanence. Only through this analogical consciousness can there be a conscious and non-moralistic demeanour of the *willing ego*, which would now be moved not through a heteronomous normative "must" (whether religious or lay), but thanks to a profound *"feel"*: that *aesthetical stance* that produces balance of the *whole* through *each one* and in which the only existential *redemption* worthy of the name is made possible.

We human beings have made many serious ideological drifts, in fact all driven by fear, and the fundamental fear is that of death. We do not know how to accept Death because we have not received and acknowledged Life. The *de-naturation* implied by a *culture* that makes itself *other* than *nature* in fact denies life, because life is *bio-logos* and is *relationship*. Relationship with the systemic and choral womb in which it is always a process that transcends individual subjects and which, of their *ensemble*, is the essence. Every time one isolates oneself from his/her system of (biological) relationships, alienation is produced: Richard Louv's *nature deficit disorder* mentioned before. Crawling along alienated existences entails never, ever feeling full of existence. The *game of substitutions*[106] takes the place, illusorily, of a fullness that can only belong to the totality, which in order to be experienced requires the decay of any target or memory (as 'conditioning', with Krishnamurti[107]) to leave room only for experience without polarisation, i.e., without observer.

In this sense, the current forms of bio-technology and mainstream medicine, which self-proclaim to aim to improve well-being and life expectancy, constitute an immense failure, not so much as prostheses or instruments of rescue and relief (functions that are fulfilled admirably and desirably), but as palliatives and patches with respect to the real problem

---

[106] (*The Dance of Life*), *La danza della Vita,* by Filippo Carli, Mauro Bergonzi, Antonia Tronti, Chapter: (*Where Life Happens*) *Dove la Vita accade,* by Antonia Tronti, page 46. *La Parola* Publisher, 2013.

[107] (*About living and dying*) *Sul vivere e sul morire,* Jiddu Krishnamurti, *Astrolabio* Publisher, 1992.

that remains untouched: that of man's *position* with respect to a hermeneutic cycle that is in fact jammed at the alienated, objectivising stage. And failure, therefore, arises insofar as these clumsy surrogates exclude from their *criterium* biological semantics and its non-invariance with respect to the system of meanings that is implemented *casu per casu* and to which the phenomenology of the living is inextricably coordinated. Disguised as scientificity/science, by only increasing the *quantity* of the existence, they contribute nothing to the enrichment of its *semantic quality*, so much so that, on the contrary, by sinking it down, they become substitutional narcotics, removers of meaning and responsibility.

And the consequences are devastating as we are reaching 9 billion 'unhappy unconsciousnesses' (paraphrasing Hegel), thrown into a systematisation that by now, in order to 'manage' them, must 'fear' them. The reckless and genocidal elite's corporations policy (not the humans in themselves) set the bases so that we are crushing the planet under our weight, under our waste, our consumption, our technical presumptions, and even more, we continue to reproduce ourselves without conscience and without choosing, rather driven only by debts and emotional imbalances or motivated by the compulsion to correspond to models of "realised persons" propagated by both lay and religious egregores which swallow millions of adepts.

No one ever reminds us to live with *death* as a 'faithful companion'. As Jacques Derrida[108] has already reminded us (with a *meditatio mortis* that is phenomenological and not narcissistic, nor metaphysical), by our side it is the greatest ally because being deeply aware, not only intellectually, but at the level of *deep perception*, that "one day you have to give way, step aside" would allow us to evaluate much more carefully how we spends every moment of our living. The *removal of death* in the society of prosthetics and anti-ageing and the *myth of eternity*, implicitly supported and deviously sold through countless more or less unintentional, subliminal, contrivances, are the most powerful tool by which one manages to induce a human being to work for 30-40 years, 8 hours a day, checking lists of papers, filling out tables, or arranging shelves, in order to secure for himself a car, a roof over his/her own head, three meals a day, two weeks' holiday in August and one at Christmas, giving to his/her beloved dears and him/herself the scraps and leftovers of his/her own energies, his/her own days, his/her own ideas, and the terminal of his/her own life years. This is the greatest crime ever conducted against the human

---

[108] *Of the spirit. Heidegger and the Question*, Jacques Derrida, Feltrinelli, Milan 1989.

being, against Life and against Nature. The consciousness of death and of the illusion of the *'part in itself'*, within the clear perception of the choral state of existence, full of care and relationship, leads exactly to the opposite of the nihilism produced by the idolatry of an individuality escaping from the *true* and the *good* by nature.

Always knowing - not believing! - that Life is made of birth and death, of becoming and of transformation with a subtle and encompassing balance. This awakened perception is the only remedy for a human who has lost the right form of *reverence* for Life-Nature. Such reverence is not based on a hypostatised 'belief' (in a creator God, other than the world, with a providential design), a perspective that is totally anthropomorphic and naive. But such demeanour and respect are based on an immanent and bio-logical *deep "I do feel"*, which has no culture, no ideology, no time, no place. So many forms and native languages, a single human language, proper of all living beings too: that of Life.

In this elegance, the technological gesture, the experiment, the research, the objectification, the manipulation are, yes, possible and licit, but because, and provided that, they are "on the Life scale" (always contextual).

If we reminded ourselves that our journey is as precious as nothing else, and limited, we would begin to demand our existences now. With the perception of death, the sacredness of life and of the true expression of ourselves would become clear again, without ever putting it off to a "later" that is too self-deferring, and the rejection of any reification would be far more lucid and firm. If technology were genuinely adopted at the service of human beings, there would immediately be an end to any repetitive and degrading "profession", which could be carried out by machines very well or be eliminated (especially insofar as it pertains to the maintenance of the colossal merry-go-round of organisations and institutions existing only as collective and falsely transcendent beliefs, since they have no ontological or, more naively, natural grounding).

No longer human beings would be told that in order to live they need someone to pay someone else and that therefore they must 'have a job', because they would know that the economy is that of resources and not that of money. Clearly, there is no point in creating economic competition: it is enough to share knowledge and resources, planning well how to preserve and renew them and simply considering 'who needs what'. Productivity since the industrial revolution has increased at least a hundredfold (speaking generally and on average), and there is no reason why everyone's 'work' cannot be decreased by at least tenfold and, above all, performed creatively. This does not at all imply a chaotic society or annihilated by inactivity, but rather a choral and profound commotion for life and deep relating to the valorisation of it, through all of us, at our best. We claim a social relationship for the (human and all living) beings and not the living beings for the society (which is *nobody*, but only a represented structure) because, as A. Camus reminded us, «one does not convince an abstraction»[109] and therefore one does not solve problems with normalising systematisations or institutions, because those are only alienated objectualities and not sentient/feeling subjects.

If we understand that in Nature there is room for Culture (as a poiesis from the former) and that they are not two different worlds, but mutually retroactive expressions of the same becoming wholeness, then we can envisage a society in which we have returned to Nature, but on a 'step above': that is why evolution exists.

The 'one step above' is that of *coherence* in which men organise themselves without any form of monetary valuation or barter, but only by sharing. All those tertiary apparatuses in which many human beings spend their lives (from finance, to law, advertising, up to any form of bureaucracy in general) would gradually decay, since they are only necessary as long as being alive is believed to be possible within the quantification and entification of supra-individual structures that find no place in nature. And such 'structures' are believed to be necessary because it is believed that a 'rule from above' must exist in order to live together; thus any poiesis of self-consciousness from below, the real condition for a natural-cultural collectivity, is sabotaged a priori.

---

[109] We quote the passage in full: «Of course, it is not the first time that men have been faced with a materially walled-off future. Usually, however, they have emerged victorious through words and screams. They appealed to other values that could provide them with hope. Today, no one speaks anymore (except those who repeat themselves). In fact, the world seems to us to be guided by blind and deaf forces that are unwilling to listen to cries of warning, nor advice, nor pleas. Something in us has been destroyed by the spectacle of the years we have just passed. And this something is that eternal trust in the human being, which always made us believe that we could receive human stimuli from the *other* (*different from oneself*) simply by speaking the language of humanity. We have seen lying, demeaning, killing, deporting, torturing, and each time it has been impossible to persuade the one who did it not to do it, because he was confident and because we cannot convince an abstraction, or the representative of an ideology. The long dialogue between men is blocked. And, of course, a man who cannot be persuaded is a frightening man. Which means that next to people who do not speak because they judge it useless, an immense conspiracy of silence has extended and is extending, accepted by those who tremble with fear and find good reasons to hide the tremor from themselves, and provoked by those who have an interest in doing so». Taken from *The Century of Fear*, an article from the series of articles *Neither Victims nor Executioners*, written by Albert Camus for the journal *Combat*, 1946.

We are inviting to turn our gaze to a *systemic*, and not *systematic*, organisation in which each subject expresses itself according to its own qualities, aptitudes and archetypes. A cosmopolitan and communitarian society that operates as a *domain of coherence*[110] and as such can in turn resonate with other social nuclei, provided all of them are on a human scale. This is the *local solution* to the *global problem* (as Coline Serreau put it[111]).

'On a human scale' means that the system must have extensions and amplitudes that can be accommodated, processed and "taken on" by the "emotional reach" and *response*-ability of each subject. To give an example: the death of a loved one is a very great sorrow, sometimes insurmountable or surmountable only after long internal processes; the death of a loved one's dear is undoubtedly a mourning and a sorrow, especially with regard to the suffering perceived by who is close to us; the death of a distant relative is a mourning, as well as of a fellow citizen of whom we had knowledge, the death of a well-known personality is a fact that, often, more than toward grief, turns us to the lost human 'capital' (artistic, intellectual, etc.), and so on, diluting. Up to the point that if we are told that 40 children per minute are dying in about 20 nations of the globe, what do we do? Are we able to take on that pain? Yet do we have the right, given that we are all responsible for everything (since we are in a relationship), to feel exempt from caring about this drama? And, conversely, if the individual human being really (emotionally) took on that 'burden', would he not collapse under such a weight?

Clearly, the above are, in practice, questions that are either inadequately posed or imply a non-unique, but dialectical answer: both 'yes' and 'no'. How to get out of the aporia? The healthiest way is by not creating the conditions to find ourselves in it. As a matter of fact, especially those who adhere to the 'western' model (today, as much in the west as in the east of the world) are responsible for almost everything that happens to the planet, and perhaps even more so for what happens to other human beings, who suffer from certain economic-political-social and technological dynamics on the 'bitter' side. But the positivised social structure – since objectual – lacks self-awareness. That structure, the *system*, is *nobody*; it cannot "feel pain", nor "feel in any way" (responsible or guilty or

---

[110] Physically meant as a set of components capable of oscillating in phase and therefore without collisions, i.e., capable of sharing a single, common 'state' because they are resonant, and therefore with an expression on necessarily shared degrees of freedom.

[111] *Solutions locales pour un désordre global*, documentary, directed and screenplay by Coline Serreau, 2010.

whatever), there is no subject that can do so. It lacks all that part which is present in everyone's conscience, deputed to "feelings", but which does not contribute at all to the constitution of the system itself, except in the ethical-moral polarisations – which are often heteronomous too – or ideological polarisations that orient some of its ranks and not others[112]. This gap, this distance between the *emotional reach* of an *intact subjectivity* and the *aseptic indifference* of an *alienated objectuality*, is what makes the events and the positivised structures, in which the individual finds himself to exist, to be out of *human's reach*/scale (out of the living scale).

An objectified structure cannot relate to the original subjects from which it was produced (and then departed) unless it objectifies them: the system treats subjects as objectified individuals, through the various representations and roles that, from time to time, act as a window of interaction with it (student, spouse, taxpayer, consumer, legal subject, believer, worker, user, citizen, etc.). In response to more or less serious problems, many of us can often only hear phrases that begin with "it is the rule which established that...", "but the law implies this or that...", "I am merely executing the requirements...", "we do not write the contracts by our own, we can just fill in them...", etc. with the increase of our "organisation", the possibility of relating among us authentically disappears, making way for just communicating and reporting.

Only living beings can care of the 'world' in their own *feeling*, but this event can only produce concrete effects if the dilution given by the anonymous objectification – which, moreover, turns out to be the condition for the very functioning of the great apparatuses that move society (markets, law plans, research programmes, institutions, etc.) – is replaced by a direct transduction of psychic, biological, perceptive (i.e.,: emotional) experiences within the collective context. In the current state of *total organisation* (i.e.,: total *Verwaltung*[113], with T.W. Adorno and M. Heidegger), the dilatation, the dilution, the impossibility of care through the emotional resource of each human being are such that there is no

---

[112] On the *fictional nature* of social and normative entities, cf. (*The Dominion of Nothing. Autopsy of supra-individual entities*) in italian, *Il Dominio del Nulla. Autopsia degli enti sovraindividuali*, Master's thesis by Andrea Porcheddu, Faculty of Philosophy, Vita-Salute San Raffaele University, Milan, AA 2011-2012. To better understand through practical examples, what it means and what the fact that the social system can no longer be a *sentient subject*, but only an *acting objectuality* due to the inexorable collapse of the sensibility proper to each human subject into the aseptic procedurality of positive institutions, see the documentary film *The Corporation*, directed by Mark Archbar and Jennifer Abbot, written by, and based on the book by, Joel Bakan, *The Corporation. The Pathological Pursuit of Profit and Power*, in particular from minute 41:00.
[113] Cf. (*Of Reality. Goals of Philosophy*) in italian, *Della realtà. Fini della filosofia*, Gianni Vattimo, Garzanti, Milan, 2012

possibility of 'feeling', there is therefore no longer the 'energy' to make the real effects of certain practices reach those structures that can decide on them, since those structures are indeed made up of persons, but they are not self-conscious beings (although they are legally entities... but we are in the realm of formal representations, not of the living dynamics).

So we are talking about the use of technical instruments (in scientific investigation, socio-economic and political organisation, and technological expression) that are always expressed within the consciousness of a *limit* (indefinable in general, but perceptible in each specific situation) that has as its criterion and guide that *sentiment* (mentioned in the introduction) of *care* for what exists and the *awareness* of being inhabitants (natives) of a precious place that welcomes us and on whose integrity and balance we all depend. We can also refer to Ivan Illich's sense of *conviviality* (as opposed to productivity), which he well emphasises: «the transition from productivity to conviviality means replacing a technical value with an ethical value, a materialised value with a realised value. Conviviality is individual freedom realised in the production relationship within a society endowed with effective instruments. When a society, whatever it may be, represses conviviality below a certain level, it becomes prey to scarcity; indeed no hypertrophy of productivity will ever succeed in satisfying the needs created and multiplied by competition»[114]. And if we speak of 'ethics' we are referring to a principle that is fundamentally ontological, bio-logical, and not ideological or moral, and that allows us not to fall into the "I *technically can,* therefore *I want* (and therefore I do)". In this sense, we join Hans Jonas in his great advancement regarding the construction of a *philosophical biology*[115], although we can now support a systemic vision on physical and neuro-biological arguments that were not available at the time when Jonas wrote and that further support the

---

[114] See (*Conviviality*) *La convivialità*, by Ivan Illich, Boroli Editore, 2005, p. 29, where this assertion comes from an equally significant consideration, which better clarifies the Illichian category of conviviality: «Each of us defines himself in his relationship with others and with the environment and by the underlying structure of the tools he uses. These tools can be ordered in a continuous series having at one extreme the dominant tool and at the opposite extreme the convivial tool: the passage from productivity to conviviality is the passage from the repetition of scarcity to the spontaneity of gift. [...] The industrial relationship is a conditioned reflex, the individual's stereotyped response to messages emitted by another user, whom he will never know, or by an artificial environment, which he will never understand; the convivial relationship, always new, is the work of people who participate in the creation of social life».

[115] Cf. H. Jonas: (*Organismus und freiheit*), *The Phenomenon of Life. Towards a Philosophical Biology*, Northwestern University Press, 2001; and (*Das Prinzip Verantwortung: Versuch einer Ethik für die technologische Zivilisation*), *The Imperative of Responsibility. In Search of an Ethics for the Technological Age,* University Chicago Press, 1985.

possibility of founding a *social feeling*, capable of that *biological ethics* we are invoking, starting from the coherence and resonance of living individuals who, emancipated from the heteronomous "I must", can self-consciously produce themselves as *nodes/links* in a *network*/web of biological relationships, in which (neuro-vegetative) *perception* is the founding process of every form of relationship and social configuration.

In all the impoverishment inherent in the current *systematisation of lives*, the mentality of a science that has become *techne* is undoubtedly a nodal motif, and it is therefore from the scientific thinking, which today discriminates every gesture of man, that the great turning point must be made, since technology now seems to be the only *subject* (but objectual!) *of history*[116]. In essence, we are inviting every thinking subject not to become accustomed to an approach that manages knowledge through a tomographic position that deludes itself that it can list a finite number of variables, properties, components, states, quantities, etc. of any system or phenomenon, and deludes itself into thinking it can arrive at the *veracity* of what exists (of which that identified 'object' is a part) through this 'way'. The actual possibilities of *knowledge* in the mode represented by current epistemology are many, but still mutilated. Because we are faced with a *terrible simplification* (in the words of P. Watzlawick[117]), which believes reality to be a sum of parts and removes such *relationship*, such underlying "*connective tissue*" only graspable intuitively over the *whole*. This then, paradoxically, makes things surreptitiously "complicated", turning Science into a quagmire of theoretical and methodological formalisations in which *punctiform specialisation* is, paradoxically, perhaps the least tragic, albeit seriously impactful outcome. Far more tragic is to observe in which way the issues of crucial importance for humanity and the entire planet, such as those discussed so far, are dealt with by 'competent insiders'. We are well aware that, as these lines unfold on the pages, the drift of the world that is staged through statistics, data, epistemological reductionism, or metaphysical postponements, continues[118]. And it also

---

[116] (*Man is antiquated*) *L'uomo è antiquato*. Vol. II, Gunther Anders, Bollati Boringhieri, 2007, cit., p. 260: «I want to say that we – and by "we" I mean the majority of our contemporaries living in industrial countries [...] – *have renounced* (or allowed ourselves to be forced to renounce) *to consider ourselves* (or nations, or classes, or humanity) *as the subjects of history*; we have dethroned ourselves (or allowed ourselves to be dethroned) and in our place we have placed other subjects of history, indeed only one other subject: *technology*».

[117] Watzlavick P., *op. cit.*

[118] Think of the 8-letter DNA (called "hachimoji") constructed by a Japanese group that is commented on as being suitable for supporting life on earth (https://www.nature.com/articles/d41586-019-00650-8), or the mice generated from the gametes of two males using the CRISPR/CAS9

continues in the echoing of publications as authoritative as they are distant from what really matters: life.

Repeating a metaphor by Martin Fleischmann[119], science today is like a tree in which the trophy of its parts (roots, trunk, branches, foliage and leaves) is proportional to how much attention, resources and time are spent on (and in) them. The form we would see is that of an immense foliage, producing ever finer and finer branched twigs, resting on a slender trunk and thin load-bearing branches. In fact, there are thousands of publications every month, in every research context... but almost all scientific work pertains to the ever more subtle variation of conditions within which parameters of the experimental setting are evaluated, and very little literature concerns turning back to consider the premises (the trunk and roots) that led to the development of the whole vision in which it discerns, distinguishes and sophisticates things (the terminal branches). Whether in the field of particle accelerators, aqueous solutions, cellular biochemistry, genetics or pharmacology, everyone is occupied with seeing what happens with that variation there, with that substitution there, and no one cares about the fact that what and how one conceives of it, both in the drafting of the experiment and in the interpretation of its results, is already inscribed within a myriad of *a priori* assumptions of which an increasing number seem to enjoy no great solidity in terms of consistency and/or explanatory capacity[120].

---

technique, although the authors 'reassure' that they have no intention of applying it to humans (https://www.sciencenews.org/article/gene-editing-creates-mice-two-biological-dads-first-time).

[119] Personally reported to the author by the late Emilio Del Giudice (Physicist, Prigogine Medal 2009) during a second rich exchange of personal reflections on the 12th December 2013 at the Physics Department of the University of Milan, Via Celoria 16, Milan.

[120] For Instance:

In Fundamental Phisics:

- Fundamental aspects on Maxwellian electrodynamics and signal propagation in vacuum [the work of David E. Rutherford at http://www.softcom.net/users/der555/; *Electromagnetic potentials without gauge transformations*, A. Chubykalo, A. Espinoza, and R. Alvarado Flores, Physica Scripta, 84 (2011) 015009 (6pp), Iop Publishing, doi:10.1088/0031-8949/84/01/015009; *The rationale of the possible existence of solenoidal electric and magnetic fields spreading without the standard retardation condition*, Andrew Chubykalo, Augusto Espinoza, Rumen Ivanov, International Journal of Engineering and Innovative Technology (IJEIT) Volume 3, Issue 5, November 2013 (https://www.researchgate.net/publication/264347089_The_rationale_of_the_possible_existence_of_solenoidal_electric_and_magnetic_fields_spreading_without_the_standard_retardation_condition), *A generalisation of classical electrodynamics for the prediction of scalar field effects*, Koen J. van Vlaenderen (arXive https://arxiv.org/abs/physics/0305098), *General Classical Electrodynamics: A new foundation of modern physics and technology*, Koen J. van Vlaenderen, Institute for Basic Research, December 201]; Jakubowski Peter, *Unified Physics*, BoD publisher, April 2017.

- on ether, relativity and Lorentz invariance: [*Deformed Spacetime – Geometrizing Interactions in Four and Five Dimensions*, Fabio Cardone, Roberto Mignani, Springer, 2007, ISBN 978-1-4020-6282-7]
- on the violability of the Planck scale as the minimum limit of uncertainty [*Disproving Heisenberg's error-disturbance relation*, Masanao Ozawa, http://arxiv.org/abs/1308.3540v1; *Violation of Heisenberg's Measurement-Disturbance Relationship by Weak Measurements*, L.A. Rozema, A. Darabi, D.H. Mahler, Alex Hayat, Y. Soudagar, and A. M. Steinberg Phys. Rev. Lett. 109, 100404 (2012); *Universally valid reformulation of the Heisenberg uncertainty principle on noise and disturbance in measurement*, Masanao Osawa, Phys. Rev. A 67, 042105-(1--6) (2003)]
- problems with the proton mass and renormalisation of vacuum energy in the standard model and the real aetiology of the holography of the universe [*Scaling Law for Organized Matter in the Universe*, N. Haramein, Bull. Am. Phys. Soc. AB006 (2001); *Quantum Gravity and the Holographic Mass*, N. Haramein, E.A. Rauscher Physical Review & Research International 3(4): 270-292, 2013].

In the study of condensed matter:
- about the real significance of coherent states on the aetiology of condensed states [*Glasses: a new view from QED coherence* M. Buzzacchi, E. Del Giudice, G. Preparata†, MITH-98/9, https://arxiv.org/abs/cond-mat/9906395 25 Jun 1999; *QED Coherence in Matter*, G. Preparata, World Scientific Publishing Co, Pte, Ltd];
- on the actual dynamics of the solution of ions in water [*QED coherence and electrolyte solutions*; E. Del Giudice, G. Preparata, M. Fleischmann; *Journal of Electroanalytical Chemistry* 482 (2000) 110-116; *Coherence in Water KT problem in living matter*, Del Giudice-Giuliani, Eur. J. Oncol. Library, vol. 5, pp 7-23. *On the "Unreasonable" Effects of ELF Magnetic Fields Upon a System of Ions*, E. Del Giudice, M.Fleischmann, G.Preparata, G.Talpo; Bioelectromagnetics 23:522-530 (2002)];
- on the role of hydrophilic surfaces in the dissociation of the water molecule and thus the pH and the possibility of extracting energy from this process [*Oxhydroelectric Effect: Oxygen Mediated Electron Current Extraction from Water by Twin Electrodes*, R. Germano, V. Tontodonato, C. Hison, D. Cirillo and F, P. Tuccinardi, Key Engineering Materials (Volume 495), http://www.scientific.net/KEM.495.100; *Coherent structures in liquid water close to hydrophilic surfaces*, Del Giudice, Vitiello, Tedeschi, Veikov, Journal of Physics: Conference Series 442 (2013) 012028; *Old and New views on the Structure of Matter and the Special case of Living Matter*; E. Del Giudice, 3rd International Workshop Dec. 2006, Journal of Physics: CS 67 (2007); *Water Respiration, the basis of the living state*, Voeikov V., Del Giudice E., WATER; A Multidisciplinary Research Journal. 1, 1 July 2009. 1, 52 – 75];
- on the coexistence of its plurality of thermodynamic phases [*Water: A medium where dissipative structures are produced by a coherent dynamics*, Marchettini, N.; Del Giudice, E.; Voeikov, V.L.; Tiezzi, E., J. Theo. Bio. 2010, 265, 511-516. *Aqueous nanostructures in water induced by electromagnetic fields emitted by EDS, a conductimetric study of fullerene, carbon nanotube Extremely Diluted Solutions*, V. Elia, L. A. Marrari, E. Napoli, Journal of Thermal Analysis Calorimetry (2012) 107:843–851; *Impact of hydrophilic surfaces on interfacial water dynamics probed with NMR spectroscopy*, Pollack G.H., Yoo, H., Paranji R., 2011. J Phys Chem Lett. 2: 532-536. *Long range forces extending from polymer surfaces*, Pollack G.H. & Zheng J.M., 2003. Phys Rev E. 68:031408.10.1103/PhysRevE.68.031408; *Surfaces and interfacial water: evidence that hydrophilic surfaces have long-range impact*; Pollack G.H., Zheng J.M., Chin W.C., Khijniak E., Khijniak E. Jr. (2006). Adv. Colloid Interface Sci. 23: 19–27. *Effect of buffers on aqueous solute-exclusion zones around ion-exchange resins*, Pollack G.H., Zheng J.M., Wexler A., 2009. J Colloid Interface Sci. 332: 511-514].

In biochemistry and biology:
- on how water (no longer understood only as a molecule, but as a collectivity of molecules) is fundamentally involved in the genesis, structure and evolution of biological processes [*Biological Coherence and Response to External Stimuli*, H. Fröhlich, Springer & Verlag, 1988; *EM Field and Spontaneous Symmetry Breaking in Biological Matter*, E. Del Giudice, S. Doglia, M. Milani,

As Kubie has well pointed out: «*we are constantly in danger of over-simplifying the problem so as to scale it down (scale it down) to mathematical treatability*» (MacKay Conferences[121]).

The inability to come to terms (literally!) with what is not totally objectifiable and identifiable, such as the subtle and endless biological relationships or the complexity of neuro-psychic processes, is accompanied by the creation of a huge and deep gap between what is repeatable-measurable and what is not, with two possible consequences at the level of praxis. Either the 'soft sciences' make a sad attempt to 'fill' this abyss by increasingly adopting a reductive scalability and tomization of their contents in a mockery of logical-physical-mathematical "customs". Or else one subscribes to a snooty and misguided distinction between science and pseudoscience, indiscriminately swallowing up in the latter everything that does not submit to the modalities of the Galilean-Popperian method because it is too subtle and too close to the impregnability of *living information*. Many phenomena, theories, techniques and approaches (especially in fields such as medicine, energy and economics) that would deserve to be seriously investigated (and enriched by the methodologies of the hard sciences) have been (and continue to be) dismissed in the

---

0550-321/86 Elsevier Science Publisher B.V. (North-Holland Publishing Division); *A Quantum Field Theoretical Approach to the Collectiv Behaviour of Biological Systems*, E. Del Giudice, S. Doglia, M. Milani, Nuclear Physics B251 [FS13] 375-400 (1985); *Coherent Quantum Electrodynamics in Living Matter*, Emilio Del Giudice, Antonella De Ninno, Martin Fleischmann, Giuliano Mengoli, Marziale Milani, Getullio Talpo, and Giuseppe Vitiello, Electromagnetic Biology and Medicine, Vol. 24, No. 3 : Pages 199-210, 2005];

- on the most important, as yet unknown to mainstream medicine, bio-communication structure in mammals, the Bong-Han network [*Feulgen Reaction Study Of Novel Threadlike Structures (Bong-han Ducts) On The Surfaces Of Mammalian Organs*, Hak-Soo Shin, Hyeon-Min Johng, Byung-Cheon Lee, Sung-Il Cho, Kyung-Soon Soh, Ku-Youn Baik, Jung-Sun Yoo, And Kwang-Sup Soh, The Anatomical Record (Part B: New Anat.) 284b:35–40, 2005, © 2005 Wiley-Liss, Inc; *The Primo Vascular System – Its role in cancer and regeneration*, Kwang-Sup Soh, Kyung A. Kang, David K. Harrison, Springer, 2012];

- on the actual role of DNA and the centrality of the physical properties it can produce depending on its shape, from which we can see how current analysis of genomic sequences destroys all these crucial features for understanding the 'genetic code' and its morphogenetic role morfogenetico [*The DNA-"Wave Biocomputer"*, Peter P. Gariaev (Pjotr Garjajev), Boris I.Birshtein, Alexander M.Iarochenko, Peter J.Marcer, George G.Tertishny, Katherine A.Leonova, Uwe Kaempf; http://www.invertone.com/WEBDOC/DNA-Garjajev-Poponin.pdf; *Principles of Linguistic-Wave Genetics*, Gariaev, P. P., Friedman, M. J., & Leonova-Gariaeva, E. A., Principles of Linguistic-Wave Genetics; DNA Decipher Journal | January 2011 | Vol. 1 | Issue 1 | pp. 011-024];

- on the real nature of what we call a 'virus' [https://en.wikipedia.org/wiki/The_Perth_Group; and the interview with Dr. S. Lanka who highlighted how the work 'proving' the existence of the measles virus pertains to experiments conducted according to rigorous practice and much more:https://formazione5lb.eu/video-5lb/201703-intervista-al-dr-stefan-lanka/].

And just to mention a few.

[121] McLuhan M., op.cit.

pseudoscience dump: from Hamer's initial discoveries constituting the real "missing link" in the whole history of medicine to low-energy nuclear reactions, from homeopathy to brain entanglement, bio-genealogy or shamanic healings, and down, always with a ridiculing smirk, to a-monetary, non-capitalist geo-economic-political models based on the real resources of the planet. Just to give a few examples.

While, on the level of theory, and from an ontological point of view, this dichotomy, which is inherently irremediable within the possibilities of logic, for which nothing else than the principle of identity and non-contradiction holds, instead of being cured by reaccepting the *great removal of the non-identical*, of the not-able-to exist in itself, of the dialectical, of the continuous becoming, of the analogical, has been patched up by both atheist-mechanist and religious-finalist idolatries.

The formers deny any possibility of self-organisation with emergent intelligence, and advocate a causal flow tending towards entropy (hence also random) as meant only diachronic, and assume the order of life as merely probabilistic, but they cannot solve the problem concerning *where* "the tension to order" towards 'disorder' comes from, and are unable to contemplate, in the course of this entropic history, how something as highly complex and organised as Life, evolution, psyche, could have been produced, believing that the 'hunt for the small' will, sooner or later, deliver the *building blocks* of existence. This position is that of epistemological reductionism, which does not see that the properties of the system can be quite different from those attributable to its elementary components and which believes that only the *micro* produces the *macro* (and not vice versa).

The latters deny that Reality is a process at one with its own *dynamic law* (at one with *what "creates it"*), they deny that there is a holonomy (given the a-spatial and a-temporal connection) that in itself already contains causes and effects, beginnings and ends, matter and consciousness[122] and that constitutes all that there is: they claim the existence of a Reality *other*, *heteronomous* and *hypostatised*, than that in which "creatures resides" and which contains the purposes, designs, instructions, prime causes, "god".

Both perspectives are superstructural and fail to recognise that Reality is articulated in an oxymoronic composition of Heraclitean opposites. Composition (*symballein*) is fundamental to *analogical prehension*, the

---

[122] Please, keep in mind that these 'lexical dualisms' are purely tools to express what is in truth unspeakable.

only truly complete/encompassing one, since it is never terminative, but circular. We hope that both science and spiritual vision are imbued with contemplative and analogical thinking, the one which grasps the co-presence between *seeker* and *sought* (for science), as well as of *creator* and *created* (for the spiritual vision). Let us quote R. Panikkar on the contemplative attitude:

> «Such an attitude is opposed to the orientation of modern civilisation, both religious and secular/lay, although I would not use the two terms in this sense, since both the secular and the religious can be sacred or profane. Indeed, it seems that our society is driven by five great incentives:
> 1. heaven, above, for believers;
> 2. history ahead, for progressives;
> 3. the work to be done, for pragmatists;
> 4. the conquest of great things, for keen ones;
> 5. the desire for success, for everyone.
>
> These five incentives are radically challenged by the contemplative spirit. In fact, contemplation emphasises the importance of the *hic et nunc*, of the *actus*, the hidden *centrum*, the inner *pax*, not the elsewhere, the later, the result, the greatness of outward gestures or the *consensus* of the majority. The first of these five aspects of contemplation challenges traditional religiosity, which is too often content to defer the true values of life to an afterlife. The second challenges the fundamental dogma of a certain secularism, which has simply transposed the ideals of religious mentality into a temporal future. The third is a type of praxis that directly overturns the core values of modern, fundamentally pan-economic society. The fourth appears as an extraneous and destructive interference of the intrinsic demands of the technological world. The fifth directly challenges the dominant anthropological conception, according to which the realisation of man necessarily presupposes the victory of one»[123].

Let the character of the renaissance and romantic *genius* return: that intellectual and wise man who knows how to vault and enthuse himself in every sphere of knowledge without any discontinuity, thus recognising the intrinsic non-partitionability of Reality starting from an undivided knowledge that is already inside him/her. Academic education must change: more interdisciplinarity is needed to rebuild a fertile eclecticism, vital to a real sharing and civilisation within which technical decision-making and equal politics are not extinguished by mutual incompetence among specialised ignorants. Thoroughgoing eclecticism (not

---

[123] Panikkar R., *op. cit.*, p. 77.

qualunquistic) will be the only 'winning answer' to the loss of competence in individuals because they lack the gnoseological tools to discuss problems beyond their acute specialisation. And if competence is lost, the 'experts of that' and the 'experts of this' are created, and the fractioning of tasks and responsibilities sees its irreversible sealing, subscription. This is how any true form of 'democracy' ceases to exist[124]. Be the people sovereign, yes... but because they make themselves worthy to be so, because they *train* (pursue a true *Bildung* path), refine and place themselves in a position of research and criticism, re-founding in every *sentient individual* what pseudo-training (especially scientific) has objectified on the many *believers* (scientifical too).

So let the *man of science* be first and foremost a contemplative because only who experiences him/herself in *contemplation* is truly *free*, since is *in the present*. And the *present* cannot be sold, measured, or promised.

---

[124] But even such form of social organisation, in its own 'model' and in its own categories, must be updated to another stage: the one in which human beings no longer organise themselves in mammoth *top-down* economic-political-social apparatuses (such as 'the states'), within which *delegation* and *representation* through third parties (elected representatives) become the only possible forms of implementation for 'democratic' action. We are well aware that these are deleterious (since they imply that 'emotional dilution' and going off the 'human scale' that we discussed above). The next step is therefore that of an autopoietic (*bottom-up*) form of community (within man's emotional and existential, biological reach) based on values such as the direct relationship with the natural realm, the preservation of resources and the liberation from the slavery of (fictitiously existing) money and of "labour" to leave room for vocational expression, permitted by a healthy and productive use of *technology* (which can thus truly be an *instrument* of human emancipation from idiotic tasks and ecologically murderous practices). This means bringing the *living* back to be the *subject of history*.

«The quadrant of evolution marks the hour of a new adaptation of the species to its environment, an adaptation that requires not the negation, but the critical overcoming of that technological universe that homo faber *has created for himself as a premise for a leap forward in his evolution but in which, for the moment, he seems to have integrated himself to the point of dissipation of his own subjectivity».*

(E. Balducci, La terra del tramonto, E.C.P. Fiesole, 1992, p. 21)

"...I believe that the fundamental idea that there are separate "things" in the universe is a creation and projection of our psychology".

(G. Bateson, A Sacred Unity, Adelphi, Milan, 1997, p. 148)

"The elders Dakotas were wise. They knew that the heart of a human being hardens if it becomes estranged from nature, they knew that the lack of a deep respect for living beings and for all that grows, soon lets also the respect for human beings die. That is why the influence of Nature, which makes young people capable of deep feelings, was an important element of their education'.

(Standing Bear, Sioux Chief, c. 1870)

Bibliography and Sources

- **Adorno Theodor W.**, *Theorie der Halbbildung, Suhrkamp Verlag Frankfurt am Main*, 1972; Italian edition edited by Giancarla Sola: *Teoria della Halbbildung*, from the series *Filosofia della formazione / 7*, directed by Mario Gennari, Edizioni il melangolo, Genoa, 2010;
- **Agosti Silvano**, *Lettere dalla Kirghisia*, L'immagine Edizioni, 2004;
- **Agosti Silvano**, *Il Genocidio Invisibile*, L'immagine Edizioni, 2008;
- **Anders Gunther**, *Man is Outdated. Vol. II*, Bollati Boringhieri, 2007;
- **Benjamin Walter**, *Angelus Novus*, translated by R. Solmi, Einaudi, 2006;
- **Benjamin Walter**, (*The Artwork in the Age of its Technical Reproducibility*) *L'opera d'arte nell'epoca della sua riproducibilità tecnica*, translation E. Filippini, preface by M. Cacciari, edited by F. Valagussa, Einaudi, 2014.
- **Bergson Henri**, *Saggio sui dati immediati della coscienza*, tr. Gisèle Bartoli, Boringhieri, Turin, 1964.
- **Bertoli Antonio**, *Psycho-Bio-Genealogy*. Le origini della malattia, Macro Edizioni, Cesena, 2010, ISBN 88-6229-156-6;
- **Biglino Mauro**, Buffa P., *Made as Human*, Uno Editori, 2018;
- **Bischof Marco**, Del Giudice Emilio, *Communication and the Emergence of Collective Behavior in Living Organisms: A Quantum Approach - Review Article*, Hindawi Publishing Corporation Molecular Biology International, Volume 2013, Article ID 987549, 19 pages, *dx.doi.org/10.1155/2013/987549*;
- **Bischof Marco**, *Introduction to Integrative Biophysics*, International Institute of Biophysics, Kapellener Str., 41472 Neuss, Published in: Popp, Fritz-Albert - Beloussov, Lev V. (eds.): *Integrative Biophysics*. Kluwer Academic Publishers, Dordrecht 2003, pp.1-115. ISBN 1-4020-1139-3;
- **Blasone M.**, Jizba M., Vitiello G., *Quantum Field Theory and its Macroscopic Manifestations*, Imperial College Press, 2011;
- **Brizhik Larissa**, Del Giudice Emilio, Jørgensen Sven E., Marchettini Nadia, Tiezzi Enzo, *The role of electromagnetic potentials in the evolutionary dynamics of ecosystems*, in Ecological Modelling 220 (2009) 1865-1869;
- **Buzzacchi M.**, Del Giudice E., Preparata G., *Glasses: a new view from QED coherence*, MITH-98/9, https://arxiv.org/abs/cond-mat/9906395 25 Jun 1999;
- **Cannon W.**, *Bodily Changes in Pain, Hunger, Fear and Rage, vol. 2*, Appleton, New York (1929).
- **Cannon W.**, *The Wisdom of the Body*, Norton, New York, (1932), Tr. it. The Wisdom of the Body, Bompiani, 1956.
- **Camus Albert**, *The Century of Fear* from *Neither Victims nor Executioners*, series of articles written for the journal *Combat*, 1946.
- **Capizzi Antonio**, *Introduction to Parmenides*, Laterza, Roma-Bari, 1995.
- **Cardone Fabio**, Mignani Roberto, *Deformed Spacetime - Geometrizing Interactions in Four and Five Dimensions*, Springer, 2007, ISBN 978-1-4020-6282-7];
- **Carli Filippo, Bergonzi Mauro, Tronti Antonia**, *The Dance of Life*, La Parola Edizioni, 2013;
- **Cattaneo Elena**, GMO, *Let's stop the anti-scientific deception'*, ricerca.repubblica.it/repubblica/archivio/repubblica/2015/10/03/fermiamo-linganno-anti-scientifico37.html;

- **Chubykalo Andrew**, Espinoza A., and Alvarado Flores R., *Electromagnetic potentials without gauge transformations*, Physica Scripta, 84 (2011) 015009 (6pp), Iop Publishing, doi:10.1088/0031-8949/84/01/015009;
- **Chubykalo Andrew**, Espinoza Augusto, Ivanov Rumen, *The rationale of the possible existence of solenoidal electric and magnetic fields spreading without the standard retardation condition*, International Journal of Engineering and Innovative Technology (IJEIT) Volume 3, Issue 5, November 2013 (link on Research Gate);
- **Clark A. J.**, *The Mode of Action of Drugs on Cells*, Edward Arnold Ed, London, UK, 1933;
- **Damasio Antonio**, *Descartes' Error: Emotion, Reason, and the Human Brain*, orig. ed: Putnam, 1994, *Descartes' Error. Emotion, Reason, and the Human Brain*, Adelphi, 1995, ISBN 978-88-459-1181-1;
- **Debord Guy**, *The Society of the Spectacle*, preface by C. Freccero and D. Strumia, Baldini & Castoldi editore, 2017;
- **Del Giudice Emilio**, Giuliani Livio, *Coherence in Water KT problem in living matter*, European Journal Oncol. Library, vol. 5, pp 7-23;
- **Del Giudice E.**, Spinetti P. R., Tedeschi A., *Water Dynamics at the Root of Metamorphosis in Living Organisms*, Water Journal 2010, 2, 566-586; doi:10.3390/w2030566;
- **Del Giudice E.**, Preparata G., Fleischmann M., *QED coherence and electrolyte solutions*; Journal of Electroanalytical Chemistry 482 (2000) 110-116;
- **Del Giudice E.**, Fleischmann M., Preparata G., Talpo G., *On the "Unreasonable" Effects of ELF Magnetic Fields Upon a System of Ions*, Bioelectromagnetics 23:522-530 (2002);
- **Del Giudice E.**, Vitiello G., Tedeschi A., Veikov V., *Coherent structures in liquid water close to hydrophilic surfaces*, Journal of Physics: Conference Series 442 (2013) 012028;
- **Del Giudice E.**, *Old and New views on the Structure of Matter and the Special case of Living Matter*; 3rd International Workshop Dec. 2006, Journal of Physics: CS 67 (2007);
- **Del Giudice E.**, De Ninno Antonella, Fleischmann Martin, Mengoli Giuliano, Milani Marziale, Talpo Getullio, and Vitiello Giuseppe, *Coherent Quantum Electrodynamics in Living Matter*; Electromagnetic Biology and Medicine, Vol. 24, No. 3: Pages 199-210, 2005;
- **Del Giudice E.**, Stefanini P., Tedeschi A., Vitiello G.; *The interplay of biomolecules and water at the origin of the active behaviour of living organisms*; Journal of Physics: Conference Series 329 (2011) 012001;
- **Del Giudice E.**, Doglia S., Milani M., A Quantum Field *Theoretical Approach to the Collectiv Behaviour of Biological Systems*; Nuclear Physics B251 [FS13] 375-400 (1985);
- **Del Giudice E.**, Vitiello G., Tedeschi A., Veikov V., *Coherent structures in liquid water close to hydrophilic surfaces*; Journal of Physics: Conference Series 442 (2013) 012028;
- **Del Giudice E.**, Tedeschi, A., Water and the autocatalysis in living matter, Electromagn. Biol.Med. 2009, 28, 46-54;
- **Del Giudice E.**, Stefanini P., *Emergence of self-organization in aqueous systems and living matter*, in Current Topics in Quantum Biology-9-17, Wydawnictwo Naukowe UAM, Poznan 2014, (http://www.nextcare.it/emergence.pdf);

- **De Ninno Antonella**, A. Congiu Castellano, E. Del Giudice, *The supramolecular structure of liquid water and quantum coherent processes in biology*, Journal of Physics: Conference Series 442 (2013) 012031 doi:10.1088/1742-6596/442/1/012031, DICE2012 IOP Publishing;
- **Derrida Jaques**, *Of the Spirit. Heidegger and the Question*, Feltrinelli, Milan 1989;
- **Ehrlich Paul**, *Receptor Theory* (http://en.wikipedia.org/wiki/Receptor_theory);
- **Elia V.**, Marrari L. A., Napoli E., *Aqueous nanostructures in water induced by electromagnetic fields emitted by EDS, a conductimetric study of fullerene, carbon nanotube Extremely Diluted Solutions*, Journal of Thermal Analysis Calorimetry (2012) 107:843-851;
- **Fröhlich Herbert**, *Biological Coherence and Response to External Stimuli*, Springer & Verlag, 1988;
- **Fukuoka Masanobu**, *The Straw Thread Revolution*, Libreria Editrice Fiorentina, Quaderni d'Ontignano, reprinted 2011. Translated from English The one-straw revolution, Rodale Editor, 1980, Original title: Shizen noho wara ippon no kakumei, Hakujuso Editor, 1975;
- **Gariaev Peter P.** (Pjotr Garjajev), Birshtein Boris I., Iarochenko Alexander M., Marcer Peter J., Tertishny George G., Leonova Katherine A., Kaempf Uwe, *The DNA-"Wave Biocomputer"*,; http://www.invertone.com/WEBDOC/DNA-Garjajev-Poponin.pdf;
- **Gariaev, P. P.**, Friedman, M. J., & Leonova-Gariaeva, E. A., *Principles of Linguistic-Wave Genetics*; DNA Decipher Journal | January 2011 | Vol. 1 | Issue 1 | pp. 011-024;
- **Germano R.**, Tontodonato V., Hison C., Cirillo D. and F, Tuccinardi P., *Oxhydroelectric Effect: Oxygen Mediated Electron Current Extraction from Water by Twin Electrodes*, Key Engineering Materials (Volume 495), http://www.scientific.net/KEM.495.100;
- **Habermas Jürgen**, *Die Zukunft der menschlichen Natur. Auf dem Weg zu einer liberalen Eugenik?*, Suhrkamp, Fankfurt am Main, 2001; translated. Italian by Leonardo Ceppa, (*The future of human nature. The risks of liberal genetics*), *Il futuro della natura umana. I rischi di una genetica liberale*, Einaudi, Turin, 2002.
- **Hamer R.G.**, *Vermächtnis einer Neuen Medizin*, Friends of Dirk, Alhaurin El Grande, (1999), Tr. it. *Testament for a New Medicine*, Friends of Dirk, Alhaurin El Grande, 2003;
- **Hamer R.G.**, *Introduction to the New Medicine*, Friends of Dirk, Alhaurin El Grande (2002);
- **Hamer R.G.**, *Krebs und alle sogenannten Krankheiten*, Friends of Dirk, Alhaurin El Grande, (2004), Tr. it. Il Cancro e tutte le cosiddette malattie, Amici di Dirk, Alhaurin El Grande, 2005;
- **Haramein Nassim**, *Scaling Law for Organised Matter in the Universe*, Bulletin. Am. Phys. Soc. AB006 (2001);
- **Haramein N.**, Rauscher Elizabeth A., *Quantum Gravity and the Holographic Mass*, Physical Review & Research International 3(4): 270-292, 2013;
- **Harrington H.** James, *Business Process Improvement: The Breakthrough Strategy for Total Quality, Productivity, and Competitiveness*, McGraw-Hill Professional Publisher, 1991. Quoted at https://www.goodreads.com/author/quotes/42617.H_James_Harrington ;
- **Hazelip Emilia**, The Art of Cultivating by Letting the Earth Do It, Free School "Emilia Hazelip", 2006, http://www.silentevolution.net/docs/Dispensa-Agricoltura-Sinergica.pdf ;

- **Heidegger Martin**, *Why Poets?* in *Sentieri interrotti* (1949), Italian translation edited by P. Chiodi, La Nuova Italia, Florence, 1968;
- **Henrard Michel**, *Comprendre sa maldie*, Ed. Amyris, Brussels 2015, Ed. It. (*Understand your Disease*) *Comprenddi la tua malattia - con le scoperte del Dottor Hamer*, Macro Edizioni, Cesena, 2015;
- **Ho Mae-Wan**, *Thermodynamics of Organisms and Sustainable Systems*, lecture at conference on Environment, Agriculture, Food, Health and Economy, World Food Day, 17 October 2007, La Sapienza University, Rome, Italy (http://www.i-sis.org.uk/ThermodynamicsOfOrganisms.php)
- **Husserl Edmund**, *Idee per una fenomenologia pura e per una filosofia fenomenologica*, edited by E. Filippini, tr. Giulio Alliney, Einaudi, turin, 1950.
- **Illich Ivan**, *La convivialità*, Boroli Editore, 2005;
- **Jakubowski Peter**, *Unified Physics*, www.naturics.eu, Printed and published by BoD, ISBN: 978-7448-0188-1, April 2017.
- **Jonas Hans**, *Organismus und freheit*), *The Phenomenon of Life. Towards a Philosophical Biology*, Northwestern University Press, 2001;
- **Jonas Hans**, (*Das Prinzip Verantwortung: Versuch einer Ethik für die technologische Zivilisation*), *The Imperative of Responsibility. In Search of an Ethics for the Technological Age*, University Chicago Press, 1985.
- **Jung Carl G.**, The Ego and the Unconscious (*Die Beziehungen zwischen dem Ich und dem Unbewussten*, 1928), translated by Arrigo Vita, with an introduction by Mario Trevi (1967) Turin: Boringhieri, 1948; ISBN 88-339-0028-2.
- **Kimbrell Andrew**, *The Fatal Harvest Reader*: The Tragedy of Industrial Agriculture, ISLAND PRESS, 2002;
- **Krishnamurti Jiddu**, *On Living and Dying*, Astrolabio Editore, 1992;
- **Laborit H.**, *In Praise of Escape*, Mondadori, Milan (1990);
- **Lipton B. H.**, *The Biology of Belief*, Penguin Books India, 2005,
- **Louv Richard**, Last Child in the Woods: Saving Our Children from Nature-Deficit Disorder, Algonquin Books, 2008;
- **Malinowski Bronislaw**, *Sex and Sexual Repression among the Savages*, Bollati Boringhieri, 5th edition, 2013;
- Malvestiti Barbara, Human Dignity after the Nice Charter. A Conceptual Analysis, Orthotes Edizioni, Naples-Salerno, 2015;
- **Mancuso Vito**, Radiophonic interview on the programme (*The whole city talks about it*) *Tutta la città ne parla*, RAI-Radio3, episode of 10 March 2017;
- **Manning Wendy D.**, Fettro Marshal Neal, Lamidi Esther, *Child Well-Being in Same-Sex Parent Families*: Review of Research Prepared for American Sociological Association Amicus Brief, Manning, W. D., Fettro, M. N., & Lamidi, E. (2014). Population Research and Policy Review, 33(4), 485-502. www.ncbi.nlm.nih.gov/pmc/articles/PMC4091994/pdf/nihms-594603.pdf;
- **Marchettini N.**, Del Giudice, E., Voeikov, V.L., Tiezzi, E., Water: A medium where dissipative structures are produced by a coherent dynamics. J. Theo. Bio. 2010, 265, 511-516;
- **Marks L.**, Same-sex parenting and children's outcomes: A closer examination of the American Psychological Association's brief on lesbian and gay parenting. Social Science Research, 41(4), 735-751 (2012);
- **Mariano Costanza**, *Quando eravamo femmine*, Sonzogno Editore, 2016;

- **Mason J.W.**, *A re-evaluation of the concept of "non-specificity" in stress theory*, in Journal of Psychiatric Research, VIII, no. 323 (1971);
- **Maturana, H.R. & Varela, F.J.**, Original title (*De maquinas y seres vivo*, 1972), *Of Machines and Living Creatures, Autopoiesis, Organization of Living* (title adapted), Porto Alegre: ArtesMédicas (1997);
- **McLuhan M.**, *Understanding Media: Extensions of Man*; MacGraw Hill: New York, NY, USA,1964;
- **Merleau-Ponty Maurice**, *Phenomenology of Perception*, edited by P.A. Rovatti, tr. A. Bonomi, Bompiani / Giunti Editori 2017 (2nd digital edition).
- **Miano Francesco**, (Günther Anders: Critics of Technology and the end of the human) *Günther Anders: critica della tecnica e fine dell'umano*, Hermeneutica, pp.199-217 2008;
- **Monod Jacques**, *Chance and Necessity*, (Transl.), p.95, Collins, London, 1972;
- **Musil Robert**, *Der Mann ohne Eigenschaften*, translated by Irene Castiglia, *The man without qualities*, Second Edition February 2016, New Compton Editor;
- **Horace**, Satires I, 1, vv. 106-107;
- **Ozawa Masanao**, *Disproving Heisenberg's error-disturbance relation*, http://arxiv.org/abs/1308.3540v1;
- **Osawa Masanao**, *Universally valid reformulation of the Heisenberg uncertainty principle on noise and disturbance in measurement*, Phys. Rev. A 67, 042105-(1--6) (2003)];
- Panikkar Raimon, *La nuova innocenza, Quaderni di ricerca*, Collana a cura del Centro di Studi Ecumenici Giovanni XXIII, pp. 77-101, Priorato di S. Egidio - Sotto il Monte - Bergamo, CENS, Cernusco s/N (MI), 1993;
- **Pert Candace B.**, *Molecules of Emotions*, Scribner Editore, 1997, Molecules of Emotions, Translation L. Perria, TEA Editore, 2005;
- **Pfister Marco**, (*Manual of Application of the Five Biological Laws (discovered by Dr. Hamer) Vol. 1, Waking up from the hypnosis of 'illness'*), *Manuale di Applicazione delle Cinque Leggi Biologiche (scoperte dal Dott. Hamer) Vol. 1, Svegliarsi dall'ipnosi della "malattia"*, Secondo Natura Editore, 2013 ISBN 978-88-95713-17-5;
- **Pluchino Alessandro**, *Time, Cosmology and Free Will*, 2011, http://www.pluchino.it/blablabla/tempo_cosmologia_liberoarbitrio.pdf;
- **Pollack G.H.**, Yoo H., Paranji R, *Impact of hydrophilic surfaces on interfacial water dynamics probed with NMR spectroscopy*, 2011. J Phys Chem Lett. 2: 532-536;
- **Pollack G.H.**, Zheng J.M., Chin W.C., Khijniak E., Khijniak E. Jr. (2006). *Surfaces and interfacial water: evidence that hydrophilic surfaces have long-range impact*; Adv. Colloid Interface Sci. 23: 19-27;
- **Pollack G.H.**, Zheng J.M., Wexler A., *Effect of buffers on aqueous solute-exclusion zones around ion-exchange resins*, 2009. J Colloid Interface Sci. 332: 511-514;
- **Pollack G.H.** & Zheng J.M., *Long range forces extending from polymer surfaces*, 2003. Phys Rev E. 68:031408.10.1103/PhysRevE.68.031408;
- **Porges Stephen W.**, (*The Polyvagal Theory. Neurophysiological foundations of emotion, attachment, communication and self-regulation*) *La Teoria Polivagale. Fondamenti neurofisiologici delle emozioni, dell'attaccamento, della comunicazione e dell'autoregolazione*, translated into Italian and edited by Vittoria Ardino, Giovanni Fioriti Editore, Rome, 2011;
- **Porges S.W.**, *The Polyvagal Theory: Phylogenetic substrates of a Social Nervous System*, International Journal of Psychophysiology 42 (2001). 123-46, Elsevier, link;

- **Porges S.W.**, *Social Engagement and Attachment: a Phylogenetic Perspective*, Annals of New York Academies of Sciences, 1008: 31-47 (2003), link;
- **Porges S.W.**, *A Review of 'The Polyvagal Theory: Neurophysiological Foundations of Emotions, Attachment, Communication, and Self-Regulation*, Journal of Couple & Relationship Therapy: Innovations in Clinical and Educational Interventions Volume 11, Issue 4, 2012, link;
- **Giuliano Preparata**, *QED Coherence in Matter*, World Scientific Publishing Co, 1995;
- **Prigogine Ilya**, *Introduction to Thermodynamics of Irreversible Processes*; Prigogine, I., C.C. Thomas, Springfield,'54; Prigogine Ilya, From Being to Becoming, Ed. W.H. Freeman and Co. S. Francisco, CA, USA, 1980;
- **Regnerus M.**, *How different are the adult children of parents who have same-sex relationships? Findings from the new family structures study.* Social Science Research, 41(4), 752-770. (2012);
- **Regnerus M.**, *Parental same-sex relationships, family instability, and subsequent life outcomes for adult children: Answering critics of the new family structures study with additional analyses.* Social Science Research, 41(6), 1367-1377. (2012).
- **Renati Paolo**, *First Elements for the Foundation of a New Paradigm in Physics*, World Futures - The Journal of New Paradigm Research, Volume 72, 2016 - Issue 1-2: Symposium on the New Science Paradigm Taylor & Francis Group - Routledge (online in August 2015) http://dx.doi.org/10.1080/02604027.2015.1018054 .
- **Renati Paolo**, *Physical Analogical Foundations of Conscious Reality*, in Analogical Con-Science Series - 1, ÌNIN Editions, Lugano, October 2016. ISBN 978-88-98497-09-6.
- **Renati Paolo**, *(On the Electrodynamic Coherence in Living Systems) Sulla coerenza del vivente nell'Elettrodinamica Quantistica*, (working title) in preparation;
- **Renati Paolo**, *Electrodynamic Coherence as a bio-chemical and physical basis for emergence of perception, semantics and adaptation in living systems*, Journal of Genetic, Molecular and Cellular Biology, 24 December 2020, Enliven Archive, ISSN: 2379-5700, http://www.enlivenarchive.org/articles/electrodynamic-coherence-as-a-bio-chemical-and-physical-basis-for-emergence-of-perception-semantics-and-adaptation-in-living-system.pdf (preprint version doi: 10.20944/preprints202011.0686.v1).
- **Renati Paolo**, *QED coherence in condensed and living matter - Theoretical frameworks, experimental results and epistemological implications*, Doctoral Thesis, discussed on 26th February 2021 in the XXXIII cycle of Ph.D. Course of Complex Systems for Physical Sciences, Life and Social-Economics Sciences, Department of Physics and Astronomy, University of Catania (IT).
- **Renati Paolo**, *Relationships and Causation in Living Matter: Reframing some Methods in Life Sciences?*, Physical Science & Biophysics Journal, September 28, 2022, Volume 6 Issue 2, ISSN: 2641-9165, Medwin Publishers, DOI: 10.23880/psbj-16000217.
- **Rozema L.A.**, Darabi A., Mahler D.H., Hayat Alex, Soudagar Y., and Steinberg A. M., *Violation of Heisenberg's Measurement-Disturbance Relationship by Weak Measurements*, Phys. Rev. Lett. 109, 100404 (2012);
- **Rutherford David E.** at http://www.softcom.net/users/der555/;
- **Sartorio Mauro**, *We are our bodies: creatures of perception designed to learn without limits*, rev.32015, Amazon, ISBN: 1508701105
- **Selye Hans**, *A syndrome produced by di-verse nocuous agents*, in Nature, CXXXVIII, no. 32 (1936);

- **Serreau Coline**, direction and screenplay of Solutions locales pour un désordre global, documentary, 2010;
- Severino Emanuele, The Essence of Nihilism, p. 263, Adelphi, 1972;
- **Shanks N**, Greek R, Greek J., Are animal models predictive for humans?, Philosophy, Ethics, and Humanities in Medicine: PEHM. 2009;4: 2, doi:10.1186/1747-5341-4-2;
- **Sheldrake Rupert**, Morphic Resonance, Inner Traditions Publisher, 2009; The Science Delusion, Rupert Shaldrake, Coronet Publisher, 2013;
- **Shin Hak-Soo**, Johng Hyeon-Min, Lee Byung-Cheon, Cho Sung-Il, Soh Kyung-Soon, Baik Ku-Youn, Yoo Jung-Sun, And Soh Kwang-Sup, *Feulgen Reaction Study of Novel Threadlike Structures (Bong-han Ducts) On The Surfaces Of Mammalian Organs*, The Anatomical Record (Part B: New Anat.) 284b:35-40, 2005, © 2005 Wiley-Liss, Inc;
- **Soh Kwang-Sup**, Kang Kyung A., Harrison David K., *The Primo Vascular System - Its role in cancer and regeneration*, Springer, 2012;
- **Spinoza B.**, *Ethics*, Bompiani, 2007;
- **Stuebe Alison**, *The Risks of Not Breastfeeding for Mothers and Infants*, Reviews in Obstetrics & Gynecology. 2009 Fall; 2(4): 222-231. PMCID: PMC2812877, https://www.ncbi.nlm.nih.gov/pmc/articles/PMC2812877/;
- **Szent-Gyorgyi A.**, *Bioenergetics*; Academic Press Inc.: New York, NY, USA, 1957;
- **Toneguzzi Danilo**, (*The Neurobiological Mechanisms of DHS*), *I meccanismi Neurobiologici della DHS*, Psiche Cervello Organo journal 1/2006, www.biophysics-research.com;
- **Toneguzzi Danilo**, (*The Biological Conflict*), *Il conflitto Biologico* Psiche Cervello Organo 1/2007, www.biophysics-research.com ;
- **Umezawa H.**, Advanced Field Theory: Micro, Macro and the Thermal Concepts, American Institute of Physics: New York, NY, USA, 1993;
- **Van der Worp H. B.**, Howells D. W., Sena E. S., Porritt M. J., Rewell S., O'Collins V., & Macleod, M. R., *Can Animal Models of Disease Reliably Inform Human Studies?* (2010) PLoS Medicine, 7(3), e1000245. http://doi.org/10.1371/journal.pmed.1000245;
- **van Vlaenderen Koen J.**, *A generalisation of classical electrodynamics for the prediction of scalar field effects*, (https://arxiv.org/abs/physics/0305098);
- **van Vlaenderen Koen J.**, *General Classical Electrodynamics: A new foundation of modern physics and technology*, Institute for Basic Research, December 2010;
- **Vattimo Gianni**, (*Of Reality. Goals of Philosophy*), *Della realtà. Fini della filosofia*, Garzanti, Milan, 2012;
- **Viano Carlo Augusto**, (*The new animism of philosophy versus science*), *Il nuovo animismo della filosofia contro la scienza*, pp. 94-103 in MicroMega Almanacco della Scienza, 5/2015;
- **Vitiello Giuseppe**, *My Double Unveiled*. John Benjamins Ed., Amsterdam, 2001;
- **Voeikov V.**, Del Giudice E., *Water Respiration, the basis of the living state*, WATER; A Multidisciplinary Research Journal. 1, 1 July 2009. 1, 52 - 75;
- **Watzlavick P.**, Weakland John H., Fisch Richard, Change. *On Formation and Problem Solving*, Astrolabium, Rome, 1974;
- **Weil Simone**, *L'enracinement. Prelude à une Declaration des Devoirs envers l'Être Humain*, trad. Italian ((*The first root. Prelude to a Declaration of Duties towards the Human Creature*), *La prima radice. Preludio a una dichiarazione dei doveri verso la creatura umana*, Milan, Edizioni di Comunità, 1954.

- **Wu Haibo**, Wang Yongsheng, Zhang Yan, Yang Mingqi, Lv Jiaxing, Liu Jun, and Zhang Yong, TALE nickase-mediated SP110 knockin endows cattle with increased resistance to tuberculosis, vol. 112 no. 13 E1530-E1539, doi: 10.1073/pnas.1421587112, http://www.pnas.org/content/112/13/E1530 ;
- **Zhang Yorke**, Lamb Brian M., Feldman Aaron W., Xiaozhou Zhou Anne, Lavergne Thomas, Li Lingjun, and Romesberg Floyd E., A semisynthetic organism engineered for the stable expansion of the genetic alphabet, PNAS, vol. 114 no. 6, 17-1322, doi: 10.1073/pnas.1616443114, http://www.pnas.org/content/early/2017/01/17/1616443114
- **Zuidgeest M.**, *The concept of matter in modern atomic theory*, Acta Biotheoretica, vol. 26, no. 1, pp. 30-38, 1977.

Other Publications in English by the Author

- *Relationships and Causation in Living Matter: Reframing Some Methods in Life Sciences?*, Physical Science & Biophysics Journal, September 28, 2022, Volume 6 Issue 2, ISSN: 2641-9165, MEDWIN PUBLISHERS, DOI: 10.23880/psbj-16000217, https://medwinpublishers.com/PSBJ/relationships-and-causation-in-living-matter-reframing-some-methods-in-life-sciences.pdf
- *Electrodynamic coherence as a bio-chemical and physical basis for emergence of perception, semantics, and adaptation in living systems*, Journal of Genetic, Molecular and Cellular Biology, 7:2020110686, 2020. ISSN 2379-5700. www.enlivenarchive.org/data/electrodynamic-coherence-as-a-biochemical-and-physical-basis-for-emergence-of-perception-semantics-and-adaptation-in-living-systems-4741.html , (doi: 10.20944/preprints202011.0686.v1).
- *1st Prize Poster presented at The Water Conference 2019 on Physics, Chemistry and Biology of Water: Temperature Dependence Analysis of the NIR Spectra of Liquid Water*, Bad Soden (Frankfurt AM, Germany), 24th-27th October 2019, presented also at The 2nd European Aquaphotomics Conference, Budapest (Hungary) December 2019;
- *Temperature Dependence Analysis of the NIR spectra of Liquid Water Confirm the Existence of Two Phases, One of Which is in a Coherent State*, P. Renati, Z. Kovacs, A. De Ninno, R. Tsenkova, Journal of Molecular Liquids 292 (2019) 111449, https://doi.org/10.1016/j.molliq.2019.111449 0167-7322/© 2019 Elsevier B.V, https://www.sciencedirect.com/science/article/abs/pii/S0167732219331617.
- *Physical Analogical Foundations of Conscious Reality*, Paolo Renati, 1st Edition October 2016, ÌNIN Editions, Centre for ÌNIN Holographic Evolving, Lugano (Switzerland), ISBN 978-88-98497-09-6; published in Italian too, by the same publisher as: Fondamenti Fisici e Analogici della Realtà Cosciente, ISBN 978-88-98497-08-9.
- *First Elements for the Foundation of a New Paradigm in Physics*, Paolo Renati, World Futures, volume 72, issue 1-2, pages 19-40, 2016 Taylor & Francis Group, Routledge, Philadelphia (published online August 2015) www.tandfonline.com/doi/abs/10.1080/02604027.2015.1018054;
- *Dysmenorrhea and endometriosis: an alternative to innovative drug therapy*, V. Corda, M. Neri, M.E. Malune, M. N. D'Alterio, V. L. Longo, M. Orrù, M. Pilloni, M. F. Marotto, P. Renati, B. Piras, A.M. Paoletti, G.B. Melis, Ass. Sandalia Solidale, Multidisciplinary Journal of Woman's Health 2015; 4(1);
- *Effectiveness of an Innovative Pulsed Electromagnetic Fields Stimulation in Healing of Untreatable Skin Ulcers in the Frail Elderly: Two Case Reports*, Fabio Guerriero, Emanuele Botarelli, Gianni Mele, Lorenzo Polo, Daniele Zoncu, Paolo Renati, Carmelo Sgarlata, Marco Rollone, Giovanni Ricevuti, Niccolò Maurizi, Matthew Francis, Mariangela

Rondanelli, Simone Perna, Davide Guido, and Piero Mannu, Hindawi Publishing Corporation Case Reports in Dermatological Medicine Volume 2015, Article ID 576580, 6 pages - www.hindawi.com/journals/cridm/2015/576580/;
- *An innovative intervention for the treatment of cognitive impairment-Emisymmetric bilateral stimulation improves cognitive functions in Alzheimer's disease and mild cognitive impairment: an open-label study.* Guerriero F., Botarelli E., Mele G., Polo L., Zoncu D., Renati P., Sgarlata C., Rollone M., Ricevuti G., Maurizi N., Francis M., Rondanelli M., Perna S., Guido D., Mannu P., Neuropsychiatric Disease Treatment. 2015 Sep 18; 11:2391-404. Doi: 10.2147/NDT.S90966. e-Collection 2015 www.dovepress.com/an-innovative-intervention-for-the-treatment-of-cognitive-impairmentnd-peer-reviewed-article-NDT;
- *Effects of electromagnetically signalized media on host-pathogen interaction,* G. D'hallewin, T. Venditti, L. Cubaiu, G. Ladu And P. Renati, in Comm. Appl. Biol. Sci., Ghent University, 79/3, 2014 - 487 in the 66th International Symposium on Crop Protection, May 20, 2014, Ghent, Belgium.

Websites:

> Gene Transfer Between Species Is Surprisingly Common. University of California, Berkeley. ScienceDaily, 11 March 2007: www.sciencedaily.com/releases/2007/03/070308220454.htm .

> Human Genome Editing - Science, Ethics, Governance, National Academy of Science - National Academy of Medicine: www.nap.edu/catalog/24623/human-genome-editing-science-ethics-and-governance

> science.sciencemag.org/content/355/6329/1040/tab-pdf

> en.wikipedia.org/wiki/The_Perth_Group

> formazione5lb.eu/video-5lb/201703-intervista-al-dr-stefan-lanka/

> www.nature.com/subjects/synthetic-biology

> www.wired.it/scienza/lab/2017/01/25/nuova-forma-vita-semi-sintetica/

> www.lescienze.it/news/2017/01/24/news/ingegnerizzare_organismi_semisintetici-3394075/?ref=nl-Le-Scienze_27-01-2017

> science.sciencemag.org/content/early/2017/03/01/science.aal1810.full

> science.sciencemag.org/content/sci/329/5987/52.full.pdf

> www.biodinamica.org/2010/09/agricoltura-industriale-colture-gm-e-biodiversita/

> it.wikipedia.org/wiki/Conigli_in_Australia

> www.lescienze.it/news/2014/09/04/news/batteri_produzione_rinnovabile_propano-2272252/

> www.wired.it/scienza/biotech/2017/07/13/film-dna-memorizzato-batterio/?utm_source=wired&utm_medium=NL&utm_campaign=daily

> www.nature.com/articles/d41586-019-00650-8

> www.sciencenews.org/article/gene-editing-creates-mice-two-biological-dads-first-time

The author

Paolo Renati, born in 1983, has a PhD in Physics of Complex Systems, with a thesis in Quantum Electrodynamics in Condensed and Living Matter, Materials Scientist Engineer, Level II Master in Materials for Micro  and Nano Technologies. For years involved in the foundation of a new paradigm in science, particularly medical and biological science, in the critical revision of the relationship between man, knowledge and technology, invoking the rediscovery of a deep feeling of Nature. Through the deepening of relevant aspects of Complexity and Quantum Field Theory applied to the study of biological matter, he has developed a vision that roots the sophisticated properties of living systems (such as semantics, memory, adaptation, perception, autopoiesis, etc.) on concrete physical grounds (such as the dynamics of coherence and resonance). From here it is possible to profoundly reframe the scenarios of major issues such as: health, environmental stewardship, how to draw nourishment and resources from it, how we humans stay and live in this world, growing up young people, and first and foremost, education in listening to the body and that subtle irreducible connection that makes us part of the web of Life.

- Ph.D. in Physics of Complex Systems (Department of Physics and Astronomy, University of Catania, Italy);
- II Level Master in Materials for Micro and Nano Technologies (IUSS, Pavia, Italy);
- M.D. in Condensed Matter Science and Engineering (Department of Sciences and Engineering, University of Genova, Italy);
- IERSED (European Institute for Scientific Research and Documentation) Scientific Committee Member (Rome, June 2014-June2016);
- International Research Fellow at The Ervin Laszlo Institute (Italy, 2014);
- Expertise and scholar in Quantum Electrodynamics of Condensed and Living Matter, epistemology and new paradigms in biology and medicine (2010-nowaday),
- Member and Teacher at The World Water Community.

www.ingramcontent.com/pod-product-compliance
Lightning Source LLC
Chambersburg PA
CBHW061319120726

48001CB00002B/586